Demanding Sustainability

"Combining socio-technical transitions, resilience and social justice perspectives into a 'just transitions framing' the authors propose an innovative five pillar framework for policymakers, decision-makers, politicians, community groups and the corporate sector that will help them productively map out their demands for sustainability."

—Julian Agyeman, *Professor Tufts University*

John Morrissey · C. Patrick Heidkamp

Demanding Sustainability

Pillars to (Re-)Build a Shared Prosperity

palgrave
macmillan

John Morrissey
Geography Department
Mary Immaculate College
University of Limerick
Limerick, Ireland

C. Patrick Heidkamp
Environment, Geography & Marine Sciences
Southern Connecticut State University
New Haven, CT, USA

ISBN 978-3-031-18957-9 ISBN 978-3-031-18958-6 (eBook)
https://doi.org/10.1007/978-3-031-18958-6

This Palgrave Macmillan imprint is published by the registered company Springer Nature Switzerland AG
The registered company address is: Gewerbestrasse 11, 6330 Cham, Switzerland

John would like to dedicate this volume to Líadain, Ailbhe and Pat - Ar scáth a chéile a mhaireann na daoine.

Patrick would like to dedicate this work to Erin, Lorelei and Milo.

Future generations are foremost in our thoughts.

Foreword

Demanding Sustainability is important reading at a time of climate emergency, when systemic, step-change in society is increasingly urgently—desperately—needed. This accessible, short, informative volume adopts a suitably measured, yet firm and pacy tone. The title is partly in homage to Antonio Guterres' 2022 address to the IPCC report when he called for everyone in society to demand change. The reader will be left in no doubt about the need for a fundamentally stronger response to climate change, one that demands radical decarbonization and a call for no one to be left behind in the process. In this regard, the book echoes the spirit and content of the UN Sustainable Development Goals—which are heavily referenced. A strength of this book that sets it apart from many such calls to action is the richness and diversity of examples. These come from the USA, Europe, and of course the lead author's native Ireland, as well as from as far afield as the antipodes.

The book provides a comprehensive introduction to ecological, economic and social dimensions of sustainability, in an accessible format, advancing five pillars. Pillars four and five are of particular note as they extend the sustainability transition topic. Pillar four reaches beyond behaviour change and advances a broader social engagement, where deep social structures are recognized that must be addressed in a transition to a just, sustainable society. Pillar five engages with resilience, drawing upon the 'bounce forward' concept where instead of resilience being seen as an ability to return to the status quo, it is rather the ability to respond to the climate emergency by installing a more just society able to withstand shocks that are to come and to pursue a more progressive agenda in the process. As joint contributors from either side of the Atlantic, Morrissey, Heidkamp, Garland and Krak are able to guide the reader through successive chapters, drawing upon events and exemplars from both the USA and Europe. This makes for an engaging, accessible and informative volume that is suited to being read in a single sitting, as much as it is to being a referred back to and built upon as readers look to read further on the many examples highlighted.

Melbourne, Australia
August 2022

Prof. Ralph Horne

Acknowledgements

John and Patrick would like to acknowledge Scott Graves, Professor of Geography, Environmental and Marine Sciences at Southern Connecticut University for his unwavering support of this project and our other transitions, sustainability and coastal projects.

John and Patrick would like to acknowledge the input of Associate Professor Alex Girard of the Art Department at Southern Connecticut State University for his outstanding graphical design work.

Many thanks to Moze Jacobs and the West Cork Doughnut Economy Network.

John and Patrick would like to thank Rachael, Naveen and the team at Palgrave Macmillan for their support and guidance throughout the development of this volume.

John wishes to acknowledge the role of the MIC Faculty of Arts Sabbatical Leave scheme which helped to facilitate the development of this volume.

The authors wish to acknowledge the permissions granted to use extracts of the following articles:

- Heidkamp, C. P., Garland, M. and Krak, L., 2021. Enacting a just and sustainable blue economy through transdisciplinary action research, The Geographical Journal, © John Wiley and Sons and used by permission. Extracts of this article are used in Chapter 1.

- Morrissey, J. (2021) Coastal communities, blue economy and the climate crisis: Framing just disruptions, The Geographical Journal, © John Wiley and Sons and used by permission. Extracts of this article are used in Chapters 6 and 7.

Contents

Notes on Contributors

Garland Michaela is a Ph.D. Student in the Department of Geography at the University of Connecticut. Michaela's research interests are focused on marine social science, sustainability, social justice and the blue economy. Michaela is currently investigating the Long Island Sound's Blue Plan and the potential inclusion of just sustainabilities as well as the current and potential blue economy for the Long Island Sound region.

Heidkamp C. Patrick is Professor of Geography in the Department of the Environment, Geography and Marine Sciences at Southern Connecticut State University, USA, and an affiliate faculty member at the University Centre of the Westfjords in Iceland. During much of the research work for this volume he was a DAAD (German Academic Exchange Service) supported Guest Professor in the Department of Geography at the University of Cologne. He is an economic geographer with research interests in sustainability transitions in the coastal zone and transdisciplinary engagement with the Blue Economy.

Krak Louie is Executive Policy Specialist for the Pennsylvania Department of Environmental Protection working on climate change, energy and air policy and regulations. He earned a master's degree in Environmental Studies and a bachelor's degree in Geography both from Southern Connecticut State University.

Morrissey John is Lecturer in Geography at Mary Immaculate College, University of Limerick, Ireland. His research is focused on issues of sustainable development, particularly sustainability transitions, low-carbon development and challenges of low-carbon economy for urban and coastal communities. John's work is informed by transdisciplinary approaches with a focus on the spatial and social evolution of transition processes. John has experience of teaching and researching sustainability issues in Ireland, Australia, New Zealand and the UK.

Abbreviations

BAU	Business as Usual
CDC-LP	Los Pinos Community Development Committee
CO_2e	Carbon Dioxide Equivalent
COP	Conference of Parties
EU27	European Union 27 Member States
GCF	Green Climate Fund
GDP	Gross Domestic Product
GHGs	Greenhouse Gases
GND	Green New Deal
GSZs	Green Sacrifice Zones
Gt	Gigatonne
HDI	Human Development Index
IPPC	Intergovernmental Panel on Climate Change
MCJ	Model of Climate Justice
NDCs	Nationally Determined Contributions
NOAA	National Oceanic and Atmospheric Administration
OA	Other African countries
OECD	Organisation for Economic Co-operation and Development
OLAC	Other Latin American and Caribbean countries
OPF	Oslo Pensjonsforsikring
PB	Planetary Boundaries
PPP	Purchasing Power Parity
PV	Photovoltaic (Solar)
R&D	Research & Development
RE	Renewable Energy
SDGs	Sustainable Development Goals

UN	United Nations
UNEP	United Nations Environment Programme
WMO	World Meteorological Organization
WWF	World Wildlife Fund

List of Figures

List of Tables

List of Boxes

CHAPTER 1

Introduction

John Morrissey, C. Patrick Heidkamp, Michaela Garland, and Louie Krak

Abstract This chapter introduces the 5 'pillars' through which to (re-) build a shared sustainable prosperity. The five pillars aim to provide a framework through which policymakers, decision-makers, politicians, community groups and the corporate sphere might begin to consider, map out and plan for just transitions in their domains. The theoretical framing combines socio-technical transitions, social justice and just transitions perspectives; the novelty of our proposed approach here is the further inclusion of resilience perspectives, to account for the shifting relations between sustainability and resilience. Our understanding of sustainability aligns with a 'strong sustainability' perspective, whereby ecological limits represent 'hard' limits to development, limits which need to be acknowledged and respected.

Keywords Pillars · Sustainable prosperity · Socio-technical transitions · Social justice · Just transitions · Strong sustainability

> Covid-19 has shaken the world. It has already led to the loss or devastation of countless lives, while many people in vital professions are working day and night to attend to the sick and stop further spread. …. it is critical to

J. Morrissey and C. P. Heidkamp, *Demanding Sustainability*,
https://doi.org/10.1007/978-3-031-18958-6_1

view this pandemic in historical context in order to avoid repeating past mistakes when we plan for the future. (Büscher et al., 2020, p. 1)

From Crisis to Opportunity

Of the many challenges to building more just sustainability, and there are indeed various and deeply troubling events that derail progress, Covid-19 has been a unique and devastating obstacle in recent times. One hundred and seventy academics of eight different Dutch universities released a short but impactful manifesto for a more sustainable development trajectory in April 2020 at the height of the first phase of the Covid-19 pandemic. The manifesto highlights the vulnerability of current development paths and emphasizes the need for thinking across longer time-horizons as well as the need for new approaches to deliver development and prosperity (Büscher et al., 2020).

Five policy proposals to re-frame economic development are forwarded, including a move away from the form of development focused on aggregate GDP growth towards an economic framework centered on redistribution; a transition to regenerative agriculture; reduction in consumption (including travel) and debt cancellation (Büscher et al., 2020). In this volume, we forward our own support for this manifesto, by incorporating the policy proposals outlined into a broadly based set of 'pillars' based on principles of shared sustainable prosperity and expanding on a triple bottom line approach to sustainability, characterized by a balanced societal focus on ecological, social and economic imperatives. The evidence on the need to do so is now overwhelming.

Report after report from the IPCC and United Nations highlights a crisis of unprecedented scale and complexity. Global GHG emissions reached 59.1 $GtCO_2$-equivalent in 2019, a record high (UNEP, 2020). As a result of pandemic-related reductions, emissions dropped by 5.4% in 2020, but levels are rapidly returning to the pre-Covid trajectory (UNEP, 2021a). CO_2 concentrations in the atmosphere were recorded at a level of 420.99 ppm for May 2022[1] (Global Monitoring Laboratory (GML) of the National Oceanic and Atmospheric Administration [NOAA] 2022), with new records emerging at an alarming rate.

A temperature rise of 2.7 °C is likely by 2100 (UNEP, 2021a), the risks of which are highlighted in an ever-increasing number of studies

[1] A peak of 422.06 ppm was recorded by NOAA for 26 April 2022.

and reports, not least the recent paper published in *Nature* on the risk of ecological cascades and positive climate feedback loops (Lenton et al., 2019) as well as the paper by Bradshaw et al. (2021) on the 'challenges of avoiding a ghastly future'. Steffen et al. (2018) warn of the risk of the planet tipping into a hothouse state, underlining the dangers of a climate significantly changed from Holocene-like conditions. The risk and urgency of the crisis point to an unambiguous state of planetary emergency (Lenton et al., 2019).

However, there are no signs that the rapid decline of carbon emissions required is a realistic prospect in the short to medium term. Policy responses to date have been wholly inadequate. Commitments under the Paris Climate Agreement, framed as Nationally Determined Contribution (NDCs), are inadequate to bridge the emissions gap in 2030[2] (UNEP, 2020). COP26 in Glasgow 2021, like many COPs before it, continued with the slow incrementalism which is now a feature of the global community's climate action. The Glasgow COP 'did not make any fundamentally new decisions' (Editorial, 2021, p. v).

Political (and academic) narratives emphasizing personal responsibility and individual behaviour change look increasingly implausible in the context of the scale of the challenge we face.[3] As articulated in the socio-technical transitions literature, wholesale systemic transformations are required to transition society towards more sustainable configurations (Geels, 2018). Increasingly, scholars are framing public engagement and behavioral change as systemic, rather than discreet, case-by-case challenges (Chilvers et al., 2018). Scholars are also becoming increasingly focused on not only the technical or political implications of such radical systemic transformation but increasingly also on the social justice implications (Sareen & Haarstad, 2018). In the transition process, 'old injustices could be reinforced, whilst new incarnations emerge if we continue to ignore the ethical implications of our policy and investment decisions' (Jenkins et al., 2017, p. 631). Problems of access to too

[2] The difference between global total GHG emissions from least-cost scenarios that keep global warming to 2 °C, 1.8 °C or 1.5 °C is defined as the Emissions Gap (for 2030) (UNEP, 2020).

[3] Even for individual consumers in those highest consuming countries (Quatar recorded 43.6 metric tonnes of CO_2 emissions per capita in 2014; The USA recorded 16.50 metric tonnes of CO_2 emissions per capita in 2014 according to the Word Bank [World Bank, 2019]), an individual carbon footprint is miniscule in terms of the global 55 billion-ton annual CO_2 emissions problem.

much energy, including waste, overconsumption and pollution as well as challenges from not having enough energy, including lack of access, under-consumption and poverty persist (Jenkins et al., 2017). While emerging economies like China and India are characterized by rapidly increasing emissions profiles, lifestyle consumption emissions of even the most well-off in these countries remain a fraction of those of wealthy elites in the Global North (Gore, 2015).

Fundamental normative questions arise in the context of this crisis, including issues of rights, equity and justice, questions on the definition of and approach to, prosperity and consideration of losses, damage and conflict (Morrissey & Heidkamp, 2022; Stern, 2016). The Presidential Memorandum of 2016 (Office of the Press Secretary, The While House, 2016) is blunt in its assessment: 'Climate change poses a significant and growing threat to national security, both at home and abroad. Climate change and its associated impacts affect economic prosperity, public health and safety, and international stability'.

Though sustainable development conceptualizations originally did frame the social domain as an equal third of the 'triple bottom line', in the years since the Brundtland Report, abbreviated framings of sustainability have largely focused on either the bio-physical environment on the one hand and/or 'development' and growth-focused narratives on the other. The end result is that the social dimension has been afforded insufficient attention in debates on sustainable development (Eizenberg & Jabareen, 2017). While basic requirements such as potable water, medication and shelter form fundamental elements of social sustainability, needs such as education, equity and fairness and employment are central to the social domain of sustainability (Vallance et al., 2011). Ideas on justice and framing of who the winners and losers of societal constructs are, are directly related to ideas on social inequality (Rawls, 1971; Sen, 2010) and form a key focus of efforts to address social sustainability in practice. Scholarship on just transitions combines concepts and insights from three strands of justice scholarship, that is energy justice, climate justice and environmental justice, together with understandings from the transitions literature (Heffron & McCauley, 2018). Heffron and McCauley (2018, p. 74) define just transition simply as a 'just' process when societies move towards an economy free of CO_2 emissions. A just transition involves a change away from fossil fuel dependency, to a low-no-carbon socio-technical regime in such a way as to ensure societal costs, benefits and risks and are distributed in 'just' ways (Sareen &

Haarstad, 2018). While this emerging research stream has offered valuable insights, how just transitions might be operationalized and realized, beyond an obvious commitment to distributional justice and procedural justice remains poorly understood. Transitions research more broadly has scope for a more meaningful and deeper engagement with justice principles (Williams & Doyon, 2019).

Demanding Sustainability

The title of this volume has been inspired in part by the words of the UN Secretary-General António Guterres in New York on 4th April 2022, on the launch of the 3rd Intergovernmental Panel on Climate Change (IPCC) report (Guterres, 2022).

> If you live in a big city, a rural area or a small island State; if you invest in the stock market; if you care about justice and our children's future; I am appealing directly to you: **demand** that renewable energy is introduced now—at speed and at scale; **demand** an end to coal-fired power; **demand** an end to all fossil fuel subsidies.

Guterres' strident language reflects the growing impatience and frustration with the slow rate of action, the lukewarm commitment to addressing the climate crisis from the political classes and above all, the pressing need to act. The frustrations of younger generations are captured in the even more strident pleas by the activist Greta Thunberg, who has appeared at COP26, Davos 2019 and the UN Climate Summit 2019 to bluntly call for authentic and real change. Her speech at Davos 2019 was a noteworthy case in point (Thunberg, 2019):

> Adults keep saying: "We owe it to the young people to give them hope." But I don't want your hope. I don't want you to be hopeful. I want you to panic. I want you to feel the fear I feel every day. And then I want you to act. I want you to act as you would in a crisis. I want you to act as if our house is on fire. Because it is.

In support of these calls from the Secretary-General, Thunberg and countless others for immediate action, we have titled our volume 'Demanding Sustainability: Pillars to (Re-)Build a Shared Prosperity', wherein we forward a series of principles on which economic and development policy for the post-Covid era should be developed. A summary version of this work has been published in Morrissey and Heidkamp (2022).

Structure of the Volume

In this volume, the theoretical framing will combine socio-technical transitions and social justice perspectives into a 'just transitions framing'; the novelty of our proposed approach here is the further inclusion of resilience perspectives, to account for the shifting relations between sustainability and resilience. By forwarding a series of 5 'pillars', we aim to provide a framework through which policymakers, decision-makers, politicians, community groups and the corporate sphere might begin to consider, map out and plan for just transitions in their domains. The five pillars are:

- An ecological prosperity (pillar one)
- A decarbonized economy (pillar two)
- A shared (cost) burden (pillar three)
- A transformative social sustainability (pillar four)
- A just resilience (pillar five)

The pillars aim to provide a framework through which policymakers, decision-makers, politicians, community groups and the corporate sphere might begin to consider, map out and plan for just transitions in their domains. Pillar one, an ecological prosperity provides the framework and boundary for all other pillars, establishing an explicitly defined space for an ecological and just prosperity into the future—one with empirically established 'hard' boundaries, based on a strong sustainability understanding of ecological stewardship and the human rights framing provided by the SDGs. Pillar two, a decarbonized economy presents us with the immediate urgency of the climate crisis. The well-being and sustainable development trajectories of future generations critically depend on our collective efforts to realize pillar two in the present decade. In support of pillar two, pillar three a shared (cost) burden and pillar four a transformative social sustainability both acknowledge the imperative for fairness and social cohesion in our responses and recognize that the social dimension is instrumentally important for transitions. Finally, pillar five a just resilience is framed in honest recognition of the hard times to come. If pillar one established the boundaries, pillar five seeks to protect those boundaries. The just disruptions ideas presented under pillar five center on the application of justice principles and social sustainability in community responses to disruptions of socio-economic and environmental systems. Collectively, these 5 pillars provide the basis for a new prosperity, developed with a firm grounding in the triple bottom line of balanced economic–ecological–societal well-being and with a clear-eyed understanding of the costs, risks and threats from business as usual (BAU) approaches. In the following chapters, the rationale and context for each pillar are outlined, including a review of the main literature debates on the issue. A 'pillar in focus' vignette is presented, showing the principles in action; these vignettes are typically of small-scale niche innovations which contain invaluable learnings for up-scale and replication and/or translation into wider or more general contexts. Descriptions of each of the five pillars are concluded with a short 'Delivering pillar x' section, outlining immediately applicable policy, economic and/or behavioural levers which might be applied to achieve the goals of the pillar in question.

Defining Pillars for Shared Sustainable Prosperity: Theoretical Framing[4]

Overview of Theoretical Framing

This work is grounded in our long-running collaborative research in the fields of socio-technical transitions and environmental economic geography. In this volume, the theoretical framing combines socio-technical transitions, social justice and just transitions perspectives; the novelty of our proposed approach here is the further inclusion of resilience perspectives, to account for the shifting relations between sustainability and resilience. Our understanding of sustainability aligns with a 'strong sustainability' perspective, whereby ecological limits represent 'hard' limits to development, limits that need to be acknowledged and respected. Grounding these disparate theoretical elements, we apply critical pragmatism as a guiding philosophy, an approach which advocates for experimentation as a means of enacting social change 'bit by bit, unevenly, by accident as well as by design' (Wood & Smith, 2008, p. 1527). Finally, a

[4] This section is based on an edited version of text first published in Heidkamp et al. (2021). Enacting a just and sustainable blue economy through transdisciplinary action research, The Geographical Journal, © John Wiley and Sons and is used by permission.

commitment to transdisciplinarity acknowledges the need for co-learning and co-production of new knowledge across the academy and civil society, in recognition of the need for radical solutions and of the 'all hands-on deck' nature of our present crises.

Strong Sustainability

In this volume, we build on the significant progress of the past decade. The UN's Sustainable Development Goals represent a blueprint to achieve a better and more sustainable future for the global community (United Nations, 2015), for instance. We agree with Coscieme et al. (2021) and Ripple et al. (2017) that the SDGs represent a global policy achievement of some significance. The pillars presented in this volume add emphasis to core aspects of the policy response to sustainability, while themselves also incorporating and overarching the SDGs.

The SDGs are designed such that no one goal is prioritized over others (UN, 2015). While on the surface, this is a laudable recognition of the balanced priority afforded by the SDGs across economic, environmental and social domains, it nonetheless presents a somewhat flat understanding of the sustainability imperatives of the twenty-first century. The nature and pace of ecological degradation, in particular, means that all of the SDGs are dependent on how successful or not measures to arrest the climate crisis are. We are now beginning to better understand that climate change makes existing poverty worse, leads to more inequality and results in more vulnerability.

In this context, this volume takes a 'strong sustainability' perspective. As argued by Bonnedahl et al. (2022, p. 150), policy framings of sustainable development since the Brundtland report have largely drawn from weak sustainability discourse, 'in which nature provides relative rather than absolute constraints'. However, in the years since Brundtland, processes of ecological modernization, including market-based solutions, regulation and new technologies have failed to address accumulating ecological crises (Haskell et al., 2021).[5] A strong sustainability perspective posits that social and economic prosperity is fundamentally reliant on a stably functioning environmental (Haskell et al., 2021; Hopwood

[5] Bonnedahl et al. (2022, p. 150) suggest that this failure is as a 'result of structural obstacles and the so-called weak sustainability discourse'.

et al., 2005). From this perspective, the protection of 'natural capital'[6] forms a central aim of sustainability efforts (Bennich et al., 2021). The volume is therefore framed in the understanding that a flourishing environmental domain, and coherent policies to protect this, form an essential underpinning for sustainability, across the SDGs.

Sustainability challenges vary according to regional and national contexts and according to differences in 'geography, governance and technology' (Nilsson et al., 2016, p. 321). Munro et al. (2017) point to the hazards of narrow 'techno-managerial' approaches which eschew more nuanced understandings of political economy dynamics in realizing SDG 7—access to energy for all, for instance. As geographers, we recognize the key importance of spatial, scalar, temporal and regional differences in the challenge to realize sustainability in practice. We look to Jackson (2006) and Garland et al. (2019) to frame our thinking on sustainability by engaging with the key geographical concepts of space and place, distance and proximity, connection and scale, relational thinking as well as power relations and governance.

Critical Pragmatism

Pragmatism provides a framework for practical experimentation, implementation and achievement in which ideas are viewed as true when they enable us to cope and complete tasks in the real world through experience and practice (Barnes, 2008; Bridge, 2008; Coaffee & Headlam, 2008; Wood & Smith, 2008). In a philosophy of science, pragmatism does not solely examine epistemologies and methodologies but also encompasses ontologies as well; in short, pragmatism is more than just a method (Marcuse, 1995, p. 259) and in its critical form provides the ontological basis for the approach taken in this volume.

Critical pragmatism adds a reflexive dimension to the practical experimental nature of pragmatism in that it first acknowledges divergent and conflicting interests as well as significant inequalities in power amongst stakeholders, and further aims to re-envision and possibly restructure stakeholder relationships (Forester, 1993). While the practical, experimental nature of pragmatism relates directly to the action research approach advocated for in this volume, the transformative nature embodied in critical pragmatist thought correlates directly to the concept

[6] 'The elements of nature that directly and indirectly produce value or benefits to people, including ecosystems, species, freshwater, land, minerals, the air and oceans, as well as natural processes and functions' (Natural Capital Committee, 2014, p. 5).

of inclusive transdisciplinarity and the aim to not only research and analyse but effect or innovate just and sustainable transformations across society. A critical pragmatist approach enables the 'collective processes of problem framing and problem solving' inherent to 'joint experimentation and social learning that directly involve the scientific and extra-scientific expertise' (Popa et al., 2015, p. 45).

Socio-Technical Transitions

In recent decades, a burgeoning body of work has developed around sustainability transitions. This discourse argues that transitioning towards more sustainable and resilient socio-ecological systems requires a fundamental and 'radical' transformation of existing socio-technical systems. The socio-technical approach to transitions highlights the co-evolution and complex interactions between industry, technology, innovation, markets, policy, culture and civil society (Geels, 2012). It further argues that policy shifts to longer-term perspectives and approaches are critical for environmental sustainability (Geels, 2012; Kemp et al., 2007).

Socio-technical transitions (STTs) are defined as the co-evolution of social and technological relationships that foster dynamic change in socio-ecological systems. For example, before shipping containers, many people worked to load and offload cargo on break bulk ships. The advent of containerization was a niche innovation that radically altered the shipping industry. This STT changed not only the initial innovative idea behind and functions of containers but also required the creation of novel ways to ship them. Container ships differed greatly from their predecessors by increasing the efficiency of transporting cargo while generating new jobs that also led to a spatial and environmental shift in the locations of ports. Along the Eastern Seaboard of the US, the spread of container terminals has driven new economic opportunity at specific harbours, leading to growing uneven spatial development (Morrissey & Heidkamp, 2018). A series of smaller New England towns including Portsmouth (New Hampshire), New London (Connecticut) and Newport (Rhode Island) exemplify this uneven spatial development. These towns experienced a period of complete urban waterfront decline, losing the majority of their shipping traffic to the harbours with container terminals (Heidkamp, 2003; Heidkamp & Lucas, 2006). This led to a spatial reorganization of ports and port activities along the Eastern Seaboard and around the world.

Thus, the initial niche innovation of the container underwent a process of maturity that resulted in uneven spatial development and livelihood opportunities (Morrissey & Heidkamp, 2018). An innovation's shift from niche to maturity is well illustrated by the multi-level perspective (MLP) transitions approach (Geels, 2012; Nykvist & Whitmarsh, 2008), in which a successful niche innovation is subsumed into the socio-technical regime, resulting in change of varying degrees, potentially leading to transformation of the socio-technical landscape itself in some cases. However, the MLP is focused on the temporal differentiation of transition processes, lacks spatial sensitivity (Coenen et al., 2012) and thus neglects the radical geographical transformations of innovation towards outcomes such as uneven development. Moreover, the spatial dimension in the SST literature is characterized by a focus on the local—that is, place as a site for socio-technical transformation (Murphy, 2015)—and an assumption of spatial homogeneity. This then leads to a lack of understanding of how policy proposals may impact current patterns of uneven development (Bridge et al., 2013). Although sustainability transitions are geographical processes that do happen in distinct locales, Truffer and Coenen (2012) highlight the need to analyse them in a regional context. Recently, Morrissey and Heidkamp (2018) developed a model based on the MLP transitions approach that highlights the importance of a regional context. In the spatially differentiated MLP, sustainability transitions are considered heterogeneous geographical processes that occur in situated and particular places rather than being universal and pervasive in nature. The model thus contributes a focus on the spatial differentiation of transition processes compared with the purely temporal view of transition processes in the original MLP model. This allows for the conceptualization of uneven development processes in the MLP and essentially eliminates the MLP's spatial naiveté (Coenen et al., 2012) as a key criticism.

Just Sustainabilities and Just Transitions

The idea of adding environmental justice concerns into sustainability and transitions frameworks is not new (see, e.g., Swilling & Annecke, 2012 or Agyeman [2013]). As Agyeman et al. (2002) point out: 'Sustainability cannot simply be a "green" or "environmental concern". A truly sustainable society is one where questions of social needs and welfare, and economic opportunity, are integrally related to environmental limits imposed by supporting ecosystems' (p. 78). Aside from living within ecosystem limits, Agyeman (2013, p. 7) outlines the following tenets for achieving a just and sustainable society: the improvement of quality of life and well-being for all; meeting the needs of both present and future generations and providing justice and equity in terms of recognition, process, procedure and outcome. These doctrines also hold true in the blue economy context, which moves away from a terra-centric conception of the economy towards a focus on the sea.

Both just transitions and just sustainabilities frameworks focus on the integration of social and environmental justice concerns into the sustainability and transitions discourse and have already been called for in the context of the Blue Economy by Garland et al. (2019), Heidkamp and Morrissey (2018) and Bennett et al. (2019), who identify the following five priorities to chart a course towards an environmentally sustainable and socially equitable Blue Economy: (1) establish a global coordinating body and develop international guidelines; (2) ensure national policies and institutions safeguard sustainability; (3) promote equitable sharing of benefits and minimization of harms; (4) employ inclusive governance and decision-making processes; and (5) engage with insights from interdisciplinary ocean science. These principles apply equally to terrestrial economies. Furthermore, they point to an increased need for research on the socio-spatial dimensions of emerging low-carbon economies that considers the importance of regions, scale mismatches in analysis and governance and power relations amongst stakeholders.

Transdisciplinarity

The term *transdisciplinarity* was simultaneously conceptualized by several scholars in the field of education to recognize disciplinary boundaries that extended beyond and greatly surpassed traditional multidisciplinary and interdisciplinary approaches. Transdisciplinarity is the ongoing process of inquiry-driven interaction, engagement of knowledge and informative action that can assist in addressing the complexities of the world (Nicolescu, 2008). As a methodology for doing science with society (van Breda & Swilling, 2019), it inspires government, academia, industry and civil society to co-learn and co-produce knowledge in real-world laboratories (Singer-Brodowski et al., 2018). Transdisciplinarity is situated in and influenced by the surrounding environment, creating a quintuple helix framework (Fig. 1.1).

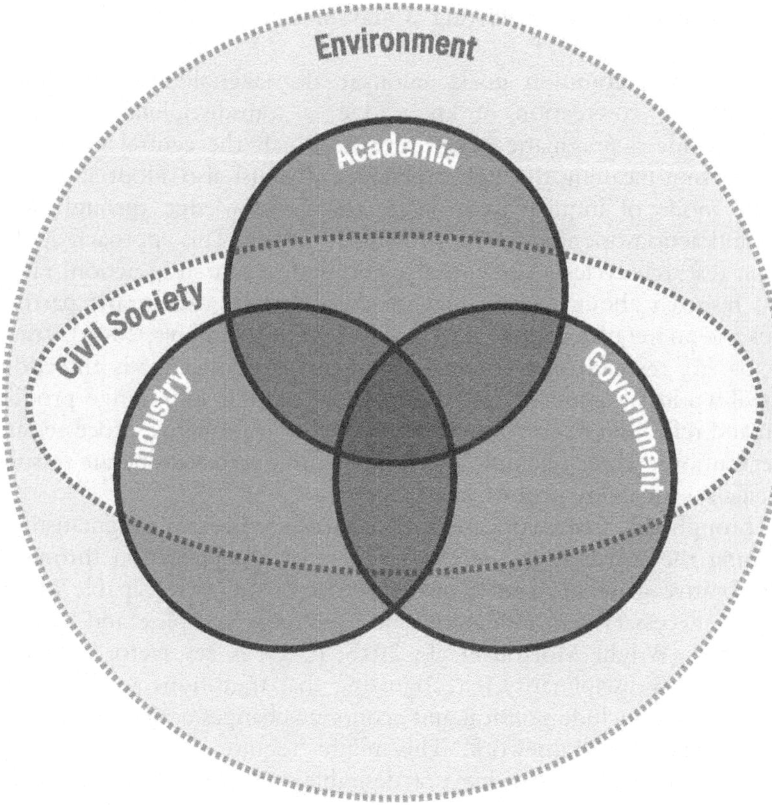

Fig. 1.1 The quintuple helix of transdisciplinarity

This structure offers an opportunity to address society's needs by requiring researchers to collaborate across academic sectors and within non-academic sectors, fostering partnership and knowledge co-production without limiting stakeholder participation. Thus, transdisciplinarity constitutes the overlap between disciplines, extends across different disciplines and transcends beyond all disciplines by seeking to understand the world through a unified framework (Nicolescu, 2008). By placing this framework in a regional context, a quintuple helix develops (Fig. 1.1), creating an agglomeration of stakeholders with socio-technical, socio-economic and socio-political goals, priorities and expectations that may be converging or conflicting (Carayannis et al., 2018; Galvao et al., 2019).

To develop common goals amongst all stakeholders through co-learning and co-creation of knowledge, a transdisciplinary approach must include a pragmatic perspective in which the central concern is constructing meaning through experience. It must also adopt abduction as the mode of inquiry to generate useful knowledge through solution and action-oriented practice (Morgan, 2014). This approach applies action research cycles (ARCs) that incorporate research in action, rather than research about action, through collaborative, democratic partnerships and an iterative approach to problem solving (Stokols, 2006). Transdisciplinary research that embraces a pragmatic perspective is embedded in real-world laboratories. ARCs allow for scientific knowledge production and reflection to be rooted in concrete, transformative processes and interventions, where scientific and non-scientific actors cooperate to solve specific sustainability problems.

Through this framework, transdisciplinarity embraces pragmatism by bridging the gap between research and practical application through a collaborative generation of knowledge that is 'uniquely capable of and perhaps necessary for addressing society's most complex and difficult problems' (Wright Morton et al., 2015, p. 1). A key factor in understanding transdisciplinarity is recognizing that transitions are long-term processes that include political and economic changes crucial to enacting a just transitions framework. This might require us to adopt long-term incrementalism to achieve a desirable outcome (Swilling, 2019). However, given the increasing urgency of the climate, biodiversity and pollution emergencies (UNEP, 2021b), the planet may not afford the time required to complete long-term projects. As a result, the TAR discourse must move beyond theoretical discussions and embrace pragmatism as a guiding principle through increased engagement with stakeholders at all levels and a transdisciplinary approach to problem-solving (Nicolescu, 2014).

References

Agyeman, J. (2013). *Introducing just sustainabilities: Policy, planning, and practice*. Zed Books Ltd.

Agyeman, J., Bullard, R. D., & Evans, B. (2002). Exploring the nexus: Bringing together sustainability, environmental justice and equity. *Sp Polity, 6*, 77–90. https://doi.org/10.1080/13562570220137907

Barnes, T. J. (2008). American pragmatism: Towards a geographical introduction. *Geoforum, 39*, 1542–1554. https://doi.org/10.1016/j.geoforum.2007.02.013

Bennett N. J., Blythe J., Cisneros-montemayor A. M., et al. (2019). Just transformations to sustainability. *Sustainability, 6*, 1–18.

Bennich, T., Belyazid, S., Stjernquist, I., et al. (2021). The bio-based economy, 2030 Agenda, and strong sustainability—A regional-scale assessment of sustainability goal interactions. *Journal of Cleaner Production, 283*, 125174. https://doi.org/10.1016/j.jclepro.2020.125174

Bonnedahl, K. J., Heikkurinen, P., & Paavola, J. (2022). Strongly sustainable development goals: Overcoming distances constraining responsible action. *Environmental Science & Policy, 129*, 150–158. https://doi.org/10.1016/j.envsci.2022.01.004

Bradshaw, C. J. A., Ehrlich, P. R., Beattie, A., et al. (2021). Underestimating the challenges of avoiding a ghastly future. *Frontiers Conservation Science, 1*, 1–10. https://doi.org/10.3389/fcosc.2020.615419

Bridge, G. (2008). City senses: On the radical possibilities of pragmatism in geography. *Geoforum, 39*, 1570–1584. https://doi.org/10.1016/j.geoforum.2007.02.004

Bridge, G., Bouzarovski, S., Bradshaw, M., & Eyre, N. (2013). Geographies of energy transition: Space, place and the low-carbon economy. *Energy Policy, 53*, 331–340. https://doi.org/10.1016/j.enpol.2012.10.066

Büscher, B., Spierenburg, M., Bavinck, M., et al. (2020). *Planning for post-corona: Five proposals to craft a radically more sustainable and equal world.*

Carayannis, E. G., Grigoroudis, E., Campbell, D. F. J., et al. (2018). The ecosystem as helix: An exploratory theory-building study of regional co-opetitive entrepreneurial ecosystems as Quadruple/Quintuple Helix Innovation Models. *R&D Managers, 48*, 148–162. https://doi.org/10.1111/radm.12300

Chilvers, J., Pallett, H., & Hargreaves, T. (2018). Ecologies of participation in socio-technical change: The case of energy system transitions. *Energy Research & Social Science, 42*, 199–210. https://doi.org/10.1016/j.erss.2018.03.020

Coaffee, J., & Headlam, N. (2008). Pragmatic localism uncovered: The search for locally contingent solutions to national reform agendas. *Geoforum, 39*, 1585–1599. https://doi.org/10.1016/j.geoforum.2007.06.010

Coenen, L., Benneworth, P., & Truffer, B. (2012). Toward a spatial perspective on sustainability transitions. *Research Policy, 41*, 968–979. https://doi.org/10.1016/j.respol.2012.02.014

Coscieme, L., Mortensen, L. F., & Donohue, I. (2021). Enhance environmental policy coherence to meet the sustainable development goals. *Journal of Cleaner Production, 296*. https://doi.org/10.1016/j.jclepro.2021.126502

Editorial. (2021). COP 26 in Glasgow. *International Journal of Refrigeration, 132*. https://doi.org/10.1016/j.ijrefrig.2021.11.029

Eizenberg, E., & Jabareen, Y. (2017). Social sustainability: A new conceptual framework. *Sustainability, 9*, 1–16. https://doi.org/10.3390/su9010068

Forester, J. (1993). *Critical theory, public policy, and planning practice: Toward critical pragmatism*. SUNY Press.

Galvao, A., Mascarenhas, C., Marques, C., et al. (2019). Triple helix and its evolution: A systematic literature review. *Journal of Science Technology Policy Management, 10*, 812–833. https://doi.org/10.1108/JSTPM-10-2018-0103

Garland, M., Axon, S., Morrissey, J., et al. (2019). The blue economy: Identifying geographic concepts and sensitivities. *Geography Compass*, 1–21. https://doi.org/10.1111/gec3.12445

Geels, F. W. (2012). A socio-technical analysis of low-carbon transitions: Introducing the multi-level perspective into transport studies. *Journal of Transport Geography, 24*, 471–482. https://doi.org/10.1016/j.jtrangeo.2012.01.021

Geels, F. W. (2018). Low-carbon transition via system reconfiguration? A socio-technical whole system analysis of passenger mobility in Great Britain (1990–2016). *Energy Research & Social Science, 46*, 86–102. https://doi.org/10.1016/j.erss.2018.07.008

Global Monitoring Laboratory (GML) of the National Oceanic and Atmospheric Administration (NOAA). (2022). Monthly average mauna loa CO2. *Carbon Cycle Greenh. Gases—Trends CO2*. https://gml.noaa.gov/ccgg/trends/mlo.html#mlo. Accessed 12 June 2022.

Gore, T. (2015). *Extreme carbon inequality: Why the Paris climate deal must put the poorest, lowest emitting and most vulnerable people first*.

Guterres, A. (2022). *Video message on the launch of the third Intergovernmental Panel on Climate Change (IPCC) report*. In United Nations, Meet Cover [Press Releases]. https://www.un.org/press/en/2022/sgsm21228.doc.htm. Accessed 5 April 2022.

Haskell, L., Bonnedahl, K. J., & Stål, H. I. (2021). Social innovation related to ecological crises: A systematic literature review and a research agenda for strong sustainability. *Journal of Cleaner Production, 325*, 129316. https://doi.org/10.1016/j.jclepro.2021.129316

Heffron, R. J., & McCauley, D. (2018). What is the 'Just Transition'? *Geoforum, 88*, 74–77. https://doi.org/10.1016/j.geoforum.2017.11.016

Heidkamp, C. P. (2003). *Revitalization, gentrification, and capital on the waterfront: An analysis of waterfront revitalization in two small port cities: Portsmouth, NH and Newport*. University of Connecticut.

Heidkamp, C. P., & Lucas, S. (2006). Finding the gentrification frontier using census data: The case of Portland, Maine. *Urban Geography, 27*, 101–125. https://doi.org/10.2747/0272-3638.27.2.101

Heidkamp, C. P., Morrissey, J. (Eds.). (2018). *Towards coastal resilience and sustainability*. Routledge.

Heidkamp, C. P., Garland, M., & Krak, L. (2021). Enacting a just and sustainable blue economy through transdisciplinary action research. *Geographical Journal*,, 1–13. https://doi.org/10.1111/geoj.12410

Hopwood, B., Mellor, M., & O'Brien, G. (2005). Sustainable development: Mapping different approaches. *Sustainable Development, 13*, 38–52. https://doi.org/10.1002/sd.244

Jackson, P. A. (2006). Thinking geographically. *Geography, 91*, 199–204. https://doi.org/10.1111/j.1745-7939.2003.tb01668.x

Jenkins, K., McCauley, D., & Forman, A. (2017). Energy justice: A policy approach. *Energy Policy, 105*, 631–634. https://doi.org/10.1016/j.enpol.2017.01.052

Kemp, R., Loorbach, D., & Rotmans, J. (2007). Transition management as a model for managing processes of co-evolution towards sustainable development. *International Journal of Sustainable Development & World Ecology, 14*, 78–91. https://doi.org/10.1080/13504500709469709

Lenton, T. M., Rockström, J., Gaffney, O., et al. (2019). Climate tipping points—Too risky to bet against. *Nature, 575*, 592–595.

Marcuse, L. (1995). Amerikanischer und Deutscher pragmatismus [American and German pragmatism]. *Zeitschrift Für Philosophische Forschung, 9*, 257–268.

Morgan, D. L. (2014). Pragmatism as a paradigm for social research. *Qualitative Inquiry, 20*, 1045–1053. https://doi.org/10.1177/1077800413513733

Morrissey, J., & Heidkamp, C. P. (2018). A transitions perspective on coastal sustainability. In C. P. Heidkamp & J. Morrissey (Eds.), *Towards coastal resilience and sustainability* (pp. 15–32). Routledge.

Morrissey, J., & Heidkamp, C. P. (2022). Sustainability after COVID: Pillars for a just transition. *Environmental Sustainability*. https://doi.org/10.1007/s42398-022-00231-y

Munro, P., van der Horst, G., & Healy, S. (2017). Energy justice for all? Rethinking sustainable development goal 7 through struggles over traditional energy practices in Sierra Leone. *Energy Policy*. https://doi.org/10.1016/j.enpol.2017.01.038

Murphy, J. T. (2015). Human geography and socio-technical transition studies: Promising intersections. *Environmental Innovation and Societal Transitions, 17*, 73–91.

Natural Capital Committee. (2014). *Towards a framework for defining and measuring change in natural capital.*

Nicolescu, B. (2008). *Transdisciplinarity–Theory and practice*. Hampton Press.

Nicolescu, B. (2014). Methodology of transdisciplinarity. *World Futures, 70*, 186–199.

Nilsson, M., Griggs, D., & Visbeck, M. (2016). Policy: Map the interactions between sustainable development goals. *Nature, 534*, 320–322. https://doi.org/10.1038/534320a

Nykvist, B., & Whitmarsh, L. (2008). A multi-level analysis of sustainable mobility transitions: Niche development in the UK and Sweden. *Technological Forecasting Social Change, 75*, 1373–1387. https://doi.org/10.1016/j.techfore.2008.05.006

Office of the Press Secretary, The While House. (2016). *Presidential memorandum—Climate change and national security. Memo.* HEADS Exec. Dep. AGENCIES. https://obamawhitehouse.archives.gov/the-press-office/2016/09/21/presidential-memorandum-climate-change-and-national-security. Accessed 1 Jan 2022.

Popa, F., Guillermin, M., & Dedeurwaerdere, T. (2015). A pragmatist approach to transdisciplinarity in sustainability research: From complex systems theory to reflexive science. *Futures, 65*, 45–56. https://doi.org/10.1016/j.futures.2014.02.002

Rawls, J. (1971). *A theory of justice*. Harvard University Press.

Ripple, W. J., Wolf, C., Newsome, T. M., et al. (2017). World Scientists' warning to humanity: A second notice. *BioScience, 67*, 1026–1028.

Sareen, S., & Haarstad, H. (2018). Bridging socio-technical and justice aspects of sustainable energy transitions. *Applied Energy, 228*, 624–632. https://doi.org/10.1016/j.apenergy.2018.06.104

Sen, A. (2010). *The idea of justice*. Penguin.

Singer-Brodowski, M., Beecroft, R., Parodi, O. (2018). Learning in real-world laboratories: A systematic impulse for discussion. *GAIA—Ecological Perspective for Science and Society, 27*, 23–27. https://doi.org/10.14512/gaia.27.S1.7

Steffen, W., Rockström, J., Richardson, K., et al. (2018). Trajectories of the earth system in the Anthropocene. *Proceedings of the National Academy of Sciences, 115*, 8252–8259. https://doi.org/10.1073/pnas.1810141115

Stern, N. (2016). *Growth and sustainability: 10 years on from the Stern review.* In: Grantham Research Institute on Climate Change and the Environment public lecture. Grantham Research Institute on Climate Change and the Environment.

Stokols, D. (2006). Toward a science of transdisciplinary action research. *American Journal of Community Psychology, 38*, 79–93. https://doi.org/10.1007/s10464-006-9060-5

Swilling, M. (2019). *The age of sustainability: Just transitions in a complex world.* Routledge.

Swilling, M., & Annecke, E. (2012). *Just transitions: Explorations of sustainability in an unfair world.* United Nations University Press.

Thunberg, G. (2019). 'Our house is on fire': Greta Thunberg, 16, urges leaders to act on climate. *The Guardian.*
Truffer, B., & Coenen, L. (2012). Environmental innovation and sustainability transitions in regional studies. *Regional Studies, 46*, 1–21. https://doi.org/10.1080/00343404.2012.646164
UNEP. (2020). *Emissions Gap Report 2020.* United Nations Environment Programme.
UNEP. (2021a). *Emissions Gap Report 2021: The heat is on—A world of climate promises not yet delivered—Executive Summary.*
UNEP. (2021b). *Making peace with nature: A scientific blueprint to tackle the climate, biodiversity and pollution emergencies.*
United Nations. (2015). *Transforming our world: The 2030 agenda for sustainable development.* A/RES/70/1. United Nations.
Vallance, S., Perkins, H. C., & Dixon, J. E. (2011). What is social sustainability? A clarification of concepts. *Geoforum, 42*, 342–348. https://doi.org/10.1016/j.geoforum.2011.01.002
van Breda, J., & Swilling, M. (2019). The guiding logics and principles for designing emergent transdisciplinary research processes: Learning experiences and reflections from a transdisciplinary urban case study in Enkanini informal settlement, South Africa. *Sustainability Science, 14*, 823–841. https://doi.org/10.1007/s11625-018-0606-x
Williams, S., & Doyon, A. (2019). Justice in energy transitions. *Environmental Innovation and Societal Transitions, 31*, 144–153. https://doi.org/10.1016/j.eist.2018.12.001
Wood, N., & Smith, S. J. (2008). Pragmatism and geography. *Geoforum, 39*, 1527–1529. https://doi.org/10.1016/j.geoforum.2008.06.003
World Bank. (2019). *CO2 emissions (metric tons per capita).* In: Data. https://data.worldbank.org/indicator/EN.ATM.CO2E.PC?name_desc=false
Wright, M. L, Eigenbrode, S. D., & Martin, T. A. (2015). Architectures of adaptive integration in large collaborative projects. *Ecology and Society, 20*(Art5). https://doi.org/10.5751/ES-07788-200405

CHAPTER 2

An Ecological Prosperity (Pillar One)

Abstract This chapter presents pillar one, an ecological prosperity, which provides the framework and boundary for all other pillars, establishing an explicitly defined space for an ecological and just prosperity into the future. Pillar one is based on empirically established 'hard' boundaries, based on a strong sustainability understanding of ecological stewardship and the human rights framing provided by the SDGs. In this chapter, we present our interpretation of Raworth's *Doughnut Economy for the Anthropocene*, whereby the doughnut is reframed not as a space within which we *should* operate in accordance with ecological and social boundaries, but rather, as a space within which we *must* operate. Consequently, this is a finite space which needs urgent recognition in policy terms, and ongoing and long-term preservation, protection and regeneration.

Keywords Ecological prosperity · Strong sustainability · Ecological stewardship · SDGs · Doughnut Economy

> *A prosperous social and economic development rests on the capacity of the biosphere to sustain it.* (Herrfahrdt-Pähle et al., 2020, p. 1)

J. Morrissey and C. P. Heidkamp, *Demanding Sustainability*,
https://doi.org/10.1007/978-3-031-18958-6_2

Resource Over/Under-Consumption

The COVID-19 pandemic presents another piece of evidence of unsustainable human activity placing extreme levels of pressure on the planet's natural life-support systems (WWF, 2018). Since 1970, WWF (2018) report a 60% decline in overall populations of mammals, birds, fish, reptiles and amphibians. At the same time, the deprivations of poverty and resource scarcity remain a reality for billions of people (UNDP, 2019). Poverty alleviation remains tied to economic growth, itself stubbornly locked-in with patterns of material resource extraction and fossil fuel dependency in particular (Morrissey & Heidkamp, 2022). Our current development model contains an 'unanswerable question': how can sufficient levels of development and poverty alleviation be delivered while simultaneously reducing overall material consumption and unsustainable levels of ecological destruction? This dilemma manifests spatially as profound inequalities in both the costs and benefits of ecological and resource exploitation—those who consume the least resources and emit the least amount of pollution live in those places projected to be most exposed to the costs of unsustainable resource use, specifically in the dangerous impacts of climate change. There is a stark and obvious imbalance in patterns of overconsumption in countries of the Global North and persistent issues of energy poverty, poor Human Development Index (HDI) performance and lack of basic infrastructure for communities in the poorest nations (Morrissey & Heidkamp, 2022).

Box 2.1: Key Data Points—Pillar One
WWF (2018) have reported a 60% decline in overall populations of mammals, birds, fish, reptiles and amphibians in the 4 decades following 1970. At the same time, 17% of children born in the year 2000 in so-called low human development countries will have died before age 20, compared with a figure of 1% in very high human development countries (UNDP, 2019).

While a neo-Malthusian perspective would point to the population side of this equation, it is clear that the extent of global resource use is far from distributed evenly across the global population. The poorest 3.5 billion people produce only 10% of total individual consumption-related emissions. These people for the most part, live in countries that are most vulnerable to climate change (Gore, 2015).

As described by Hubacek et al. (2017), increasing income leads to increasing carbon footprints highlighting the central challenge of achieving both climate change mitigation and adequate development to sufficiently deliver on poverty alleviation. The countries with the highest Human Development Index (HDI) scores also contribute most, in per capita terms, to the GHG emissions driving climate change (Hickel, 2020). Of further concern, UNDP report that 'under the shadow of the climate crisis and sweeping technological change, inequalities in human development are taking new forms in the twenty-first century' (UNDP, 2019, p. 3). We are therefore faced with a multifaceted challenge, a problem of overconsumption in the wealthy advanced economies combined with persistent poverty and lack of basic infrastructure in the poorest countries.

Inequality dynamics within countries are also changing, with income inequality having increased in nearly all countries in the past couple of decades, albeit at different speeds (Alvaredo et al., 2018).

Defining the Limits

The problems of resource over/under-consumption represent a foundational problem for just transitions. As argued from many years in the sustainability literature, resource use needs to be constrained within clearly defined limits. However, from a social sustainability perspective, all communities should have access to the basics for a good life; energy consumption remains fundamental to economic functioning and social prosperity (Wu & Chen, 2017). There is therefore a lower limit to resource consumption which also needs to be considered. Spangenberg (2002) used the idea of environmental space to articulate these upper and lower limits. At the upper limit of resource use, energy consumption, material flows and land use need to be managed within ecological limits. At the lower limit, a socially defined minimum of resource availability can allow individuals and communities to 'lead a dignified life' (Spangenberg, 2002, p. 295). More recent work has more precisely mapped what might be described as the ceiling(s) to resource use, which considers global carrying capacity and planetary limits.

The planetary boundaries concept developed by Rockström et al. present 9 environmental parameters, highlighting in particular those parameters where the so-called safe operating space has been breached by human activities (Rockström et al., 2009; Steffen et al., 2015). The planetary boundaries concept gives an empirical, clearly defined and readily understandable framework for establishing a ceiling on resource use across multiple parameters. The 'safe operating space' is a useful tool to identify and chart the limits of resource use and emissions levels and accordingly, to develop appropriate policy measures and responses (Hjalsted et al., 2021; Rockström et al., 2009; Steffen et al., 2015). For real-world impact, the boundaries need to be appropriately developed into performance targets that are relevant across decision-making scales (Häyhä et al., 2016).

As described, however, for much of the world's population, the issue is not one of over use of resources but rather, a lack of access to resources (Rockström et al., 2009).

Spangenberg's social floor has been represented elsewhere. An internationally agreed basis for a 'social foundation' is provided by the Sustainable Development Goals (SDGs), established in 2015. The 17 interrelated SDGs address challenges related to poverty, climate change, inequality, peace and justice and environmental degradation. These are presented in Table 2.1.

Table 2.1 The sustainable development goals

Goal No.	*Description*
Goal 1	No Poverty
Goal 2	Zero Hunger
Goal 3	Good Health and Well-being
Goal 4	Quality Education
Goal 5	Gender Equality
Goal 6	Clean Water and Sanitation
Goal 7	Affordable and Clean Energy
Goal 8	Decent Work and Economic Growth
Goal 9	Industry, Innovation and Infrastructure
Goal 10	Reduced Inequality
Goal 11	Sustainable Cities and Communities
Goal 12	Responsible Consumption and Production
Goal 13	Climate Action
Goal 14	Life Below Water
Goal 15	Life on Land
Goal 16	Peace and Justice Strong Institutions
Goal 17	Partnerships to achieve the Goal

The UN forward the SDGs as a blueprint to achieve a better and more sustainable future for the global community (United Nations, 2015). Such a social foundation provides a minimum standard of resource use below which shortfalls in well-being including hunger ill-health illiteracy and energy poverty are deemed unacceptable.

An inventive and useful combination of both the planetary boundaries and the SDGs is provided by Raworth's 'Doughnut for the Anthropocene' (Raworth, 2017). The 'doughnut' in question provides an elegant integration of two sets of boundaries. Resource and pollution limits are determined by an ecological 'ceiling', represented by the planetary boundaries work of Rockström et al. (2009) and Steffen et al. (2015). The minimum standards for social functioning are determined by a social 'floor' or foundation, which is informed by the SDGs. The doughnut is therefore comprised of two concentric radar charts, which form the social and environmental or planetary boundaries that together provide a basis for human well-being (Raworth, 2017). From a decision-making perspective, the doughnut might be applied by national regional and local policy makers to inform, frame and steer debate on sustainability transitions by empirically defining and succinctly communicating 'an ecologically safe and socially just space' for societal prosperity (Raworth, 2017, p. e48). Figure 2.1 shows our representation of the doughnut economy, based on Raworth's original.

The slight adjustment made in Fig. 2.1 is that ecological overshot and social under-provision for key parameters are represented as a 'bite out of the doughnut'. In the original, ecological overshoots and social deficits are represented as red indicators, appearing beyond the doughnut; the doughnut itself being the space within which humanity should operate 'to meet the needs of all within the means of the planet'. We have purposefully shown ecological overshoot and social under-provision as an erosion of 'ecological prosperity' to better represent the existential nature of current threats; climate crisis, biodiversity loss and threats to peace and justice not only erode current levels of well-being and prosperity, they actively erode the space for an ecological and just prosperity into the future, potentially in non-reversible ways. The challenge is therefore to steer development to the ecologically safe and socially just space of the doughnut economy in the short-medium term, so that this safe and socially just space remains a viability for future generations. The doughnut is therefore reframed not as a space within which we *should* operate in accordance with ecological and social boundaries, but rather, as a space

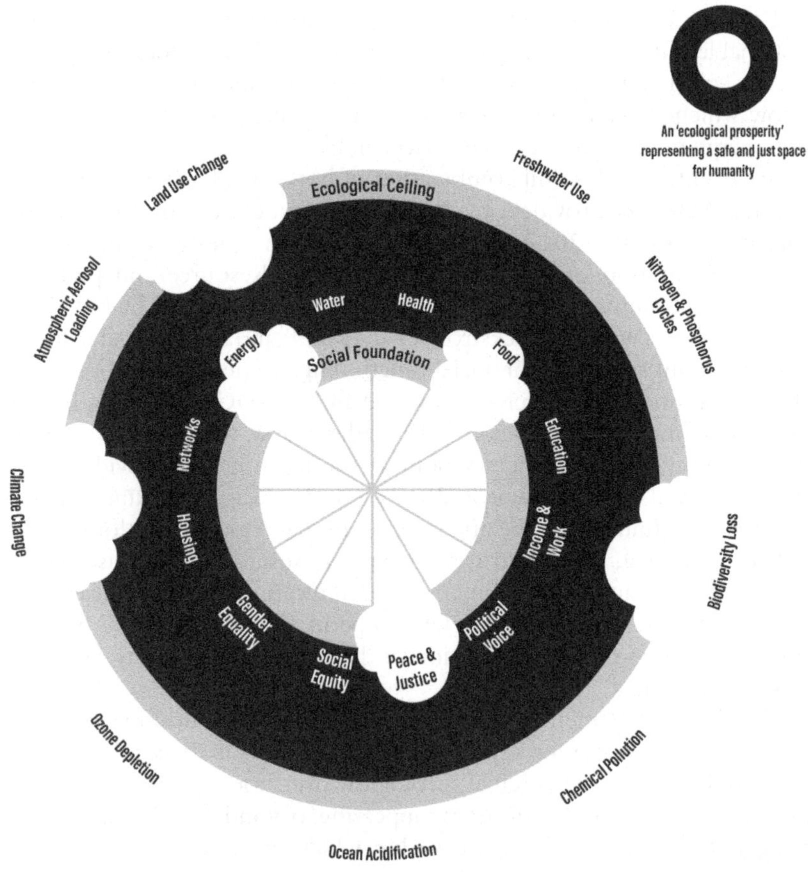

Fig. 2.1 'Eating the doughnut', after Raworth's Doughnut Economy

within which we *must* operate for a sustainable prosperity. Consequently, this is a finite space that needs urgent recognition in policy terms, and ongoing and long-term preservation, protection and regeneration.

Sustainable Development Goals

The potential for the SDGs to guide responses to the challenge of planetary boundaries is flagged by Häyhä et al. (2016, p. 69), 'the 2030 Agenda ... provides a facilitating context for the governance of the planetary boundaries. It places the (planetary boundaries) concept in the broader context of sustainable development'. Difficult policy decisions lie ahead, however. It is unclear how progress on all SDGs can be maintained in the context of rapid and deep emissions cuts, as would be required to limit global heating to 1.5 °C. The IPCC themselves flag the potential for trade-offs across sustainable development dimensions in the context of rapid decarbonization (IPCC, 2018). For policy makers, the challenges of curtailing the global economy to within the limits of the planetary boundaries, while simultaneous establishing necessary social foundations globally are immense. The IPPC (2018, p. 19) points to the likelihood of 'multiple trade-offs across the Sustainable Development Goals (SDGs)', while also highlighting some potential synergies.

The SGDs have been designed with a level of interconnectedness in mind, whereby no one goal should be prioritized over others (United Nations Department of Economic & Social Affairs, 2021). This is important for synergies across and within the goals. However, this also presents challenges of trade-offs, which increase the complexity of the goals. Linkages between climate and development were written into both the Sustainable Development Goals (SDGs)[1] and the Paris Agreement (Gomez-Echeverri, 2018). However, the Paris Agreement and the 2030 Agenda have to date been largely addressed as separate policy domains and interlinkages have received limited attention (Iacobuta et al., 2022). While 'climate change itself will have substantial negative implications for development' (Bandari et al., 2022, p. 9), mis-alignment between SGDs and climate action risks the production of trade-offs and high costs, especially for Global South countries (Gomez-Echeverri, 2018). Trade-offs in SDG realization make coherent political solutions challenging (Coscieme et al., 2021). At the national level, political aspects such as institutional barriers and organizational mandates might impede collaboration and synergies when implementing the SDGs (Bandari et al., 2022). Bandari et al. (2022)'s study of SDG implementation in the Goulburn-Murray region in Victoria, Australia, shows how the local context means

[1] 2030 Agenda for Sustainable Development.

an inevitable prioritization of certain SDGs over others, characterized by locally specific trade-offs and challenges. The means through which individual SDGs are realized also has implications for wider sustainability and for potential trade-offs between SDGs. For example, Nilsson et al. (2016) point to SDG 7 *access to energy* and the potential of coal as a means to realize this. However, such a strategy would necessarily increase air pollution (SDG 3), exacerbate climate change (SDG 13) and result in further ocean acidification (SDG 14). Negative interactions of SDG policies may also accrue as a result of institutional and governance weakness (Nilsson et al., 2016). To minimize trade-offs and maximize synergies, therefore, climate mitigation policies need to be coordinated and integrated across sectors and policy areas and governance approaches (Cohen et al., 2021; Pardoe et al., 2018). Conversely, synergies are possible. Sarkodie (2022) argues that sustained economic growth with reduced income inequality improves energy sustainability in developing economies. For climate finance, synergies between climate protection related investments and SDGs initiatives can generate incentives for donors and improve outcomes (Gomez-Echeverri, 2018).

Box 2.2: An Ecological Prosperity in Action
AN ECOLOGICAL PROSPERITY IN ACTION:
Doughnut Economy Network, West Cork, Ireland

The West Cork Doughnut Economy Network was officially launched on 24th February 2022 (West Cork Doughnut Economy Network, 2022). The Network is located in the South West region of Ireland in the West Cork Municipal District, comprised of nine small towns, numerous villages and a multitude of dispersed rural communities (Cork County Council, 2019). The online launch of the network followed from close to two years of conversations with community members together with topic-based and place-based meetings, including engagement with governance and business stakeholders (Jacobs, 2022). The network is engaged with the Doughnut Economics Action Lab (DEAL), a Community Interest Company founded in 2019 with the aim 'to help create twenty-first century economies that are regenerative and distributive by design, so that they can meet the needs of all people within the means of the living planet', fundamentally grounded in Raworth's Doughnut Economics concepts (Doughnut Economics Action Lab, 2022).

*Look at every product we buy and see if it harms the planet. If we lose biodiversity, we will end up in wasteland. We need to be able to buy healthy food locally and at cheaper prices. The problem is not nature, the problem is how people interact with nature. We need to adopt the slogan "Think Global Act Local". **Moze Jacobs, Co-founder West Cork Doughnut Economy Network**,* quoted in McCarthy (2022)

The West Cork Doughnut Economy Network aims to use the expertise and tools provided by DEAL as a point of departure for a series of deeper conversations on the meaning of prosperity, on regional priorities for positive change, on personal reflections on human–environment interactions and on focusing personal and community responses (Jacobs, 2022). The 'scaffolding' provided by DEAL tools has been important in the earlier stages of network development; the DEAL concept aims to co-create community-level solutions to help realize Doughnut Economics, through a process of creativity, experimentation, demonstration and inspiration (Doughnut Economics Action Lab, 2022).

Moze Jacobs has described the positive impact of this collaboration as providing status and substance to community discussions as well as providing inspiration on directions for network activity and development (McCarthy, 2022). The networking elements also form a core motivator for the organization, targeting collaboration and collective action through connecting individuals and groups in the West Cork Region motivated by environmental and social sustainability and community focused economies.

'Doughnut Economics intertwines with these existing groups, all working to prevent climate chaos with genuine people who all want the same thing. People interested in community often also interested in the environment' Moze Jacobs. More broadly the DEAL model relies on a central digital platform or facilitating infrastructure, around which diverse local networks attempt to operationalize Doughnut Economics in reality, on the ground, in their communities and in ways aligned with local and regional priorities (Gross, 2020). The analysis by Gross (2020) suggests that DEAL exhibits clear aspects of 'a community of transformation', typified by a core transformational philosophy concerned with the creation of novel innovations and practices.

Delivering Pillar One

While the rationale and logic of both the planetary boundaries and the SDGs might seem to be irrefutable, translating these concepts into real world and actionable policy presents a far more challenging prospect. Pillar one necessarily requires a range of policy measures targeted on ***constraint***: that is, limiting resource use in clearly targeted ways to maintain the safe operating space for humanity, now and into the future. In an economic system configured for growth, this presents a wicked dilemma, made all the starker by the challenges of poverty alleviation for large swathes of the global population.[2] Assuming the appetite exists across society, including across political, industrial and civil society stakeholders to actually deliver on the safe operating space, the following broadly defined solutions might be applied:

- Establishment of standardized international and national accounting and reporting standards for all constituent elements of the 'doughnut'
- As with the Paris Agreement for climate, internationally agreed performance targets need to be established for all of the planetary boundaries, starting with those currently transgressing the 'safe operating space'
- Official recognition at the national level of both the planetary boundaries and the SDGs as headline parameters for societal performance and prosperity
- National, regional and local dialogues on the direction of societal development for prosperity and resilience, framed by the empirical evidence of both the planetary boundaries and the SDGs
- Headline indicators for key parameters of the 'doughnut' selected as foci for national level discourse on sustainability/prosperity performance. This occurs with GDP in the economic sphere, but could as easily happen with carbon footprint, for instance
- Socio-cultural shift from a focus on unconstrained consumerism to a focus on sufficiency, resilience and ecological restoration

[2] And in the context of rapidly increasing cost of living pressures and the associated political significance of these.

- Clear policy commitment to circular economy and the absolute decoupling of economic performance from growth in material/resource consumption.

References

Alvaredo, F., Chancel, L., Piketty, T., et al. (2018). *World Inequality Report 2018*. World Inequality Lab.

Bandari, R., Moallemi, E. A., Lester, R. E., et al. (2022). Prioritising sustainable development goals, characterising interactions, and identifying solutions for local sustainability. *Environmental Science & Policy, 127*, 325–336. https://doi.org/10.1016/j.envsci.2021.09.016

Cohen, B., Cowie, A., Babiker, M., et al. (2021). Co-benefits and trade-offs of climate change mitigation actions and the Sustainable Development Goals. *Sustainable Production and Consumption, 26*, 805–813. https://doi.org/10.1016/j.spc.2020.12.034

Cork County Council. (2019). West Cork municipal district. In *Cork Local Area Plans*. http://corklocalareaplans.com/west-cork-municipal-district/. Accessed 2 Jan 2022.

Coscieme, L., Mortensen, L. F., & Donohue, I. (2021). Enhance environmental policy coherence to meet the Sustainable Development Goals. *Journal of Cleaner Production, 296*, 126502. https://doi.org/10.1016/j.jclepro.2021.126502

Doughnut Economics Action Lab. (2022). *About DEAL*. https://doughnuteconomics.org/about. Accessed 2 Feb 2022.

Gomez-Echeverri, L. (2018). Climate and development: Enhancing impact through stronger linkages in the implementation of the Paris Agreement and the Sustainable Development Goals (SDGs). *Philosophical Transactions of the Royal Society A—Mathematical Physical and Engineering Sciences, 376*,. https://doi.org/10.1098/rsta.2016.0444

Gore, T. (2015). *Extreme carbon inequality: Why the Paris climate deal must put the poorest, lowest emitting and most vulnerable people first*. Oxfam.

Gross, P. L. (2020). Better together? How the Doughnut Economics Action Lab organizes communities for transformative action. https://www.academia.edu/download/65888495/Gross_2021_DEAL_CommunityOrganizing.pdf. Accessed 4 May 2022.

Häyhä, T., Lucas, P. L., Van, V. D. P., et al. (2016). From planetary boundaries to national fair shares of the global safe operating space—How can the scales be bridged? *Global Environmental Change, 40*, 60–72. https://doi.org/10.1016/j.gloenvcha.2016.06.008

Herrfahrdt-Pähle, E., Schlüter, M., Olsson, P., et al. (2020). Sustainability transformations: Socio-political shocks as opportunities for governance transitions. *Global Environmental Change, 63*, 102097. https://doi.org/10.1016/j.gloenvcha.2020.102097

Hickel, J. (2020). The sustainable development index: Measuring the ecological efficiency of human development in the anthropocene. *Ecological Economics, 167*, 1–10. https://doi.org/10.1016/j.ecolecon.2019.05.011

Hjalsted, A. W., Olsen, K. H., Laurent, A., et al. (2021). Sharing the safe operating space. Exploring ethical allocation principles to operationalize the planetary boundaries and assess absolute sustainability at individual and industrial sector levels. *Journal of Industrial Ecology, 25*, 6–19. https://doi.org/10.1111/jiec.13050

Hubacek, K., Baiocchi, G., Feng, K., et al. (2017). Global carbon inequality. *Energy, Ecology and Environment, 2*, 361–369. https://doi.org/10.1007/s40974-017-0072-9

Iacobuta, G. I., Brandi, C., & Dzebo, A., et al. (2022). Aligning climate and sustainable development finance through an SDG lens. The role of development assistance in implementing the Paris Agreement. *Global Environmental Change, 74*, e102509. https://doi.org/10.1016/j.gloenvcha.2022.102509

IPCC. (2018). Summary for policy makers. In V. Masson-Delmotte, P. Zhai, & H.-O. Pörtner, et al. (Eds.), *Global warming of 1.5°C. An IPCC Special Report on the impacts of global warming of 1.5°C above pre-industrial levels and related global greenhouse gas emission pathways, in the context of strengthening the global response to the threat of climate change.*

Jacobs, M. (2022). *Launch of the West Cork Doughnut Economy Network.* Doughnut Economics Action Lab. https://doughnuteconomics.org/stories/154. Accessed 12 Jun 2022.

McCarthy, M. R. (2022). *Cork group's mission for a greener planet*. Echo.

Morrissey, J., & Heidkamp, C. P. (2022). Sustainability after COVID: Pillars for a just transition. *Environmental Sustainability*. https://doi.org/10.1007/s42398-022-00231-y

Nilsson, M., Griggs, D., & Visbeck, M. (2016). Policy: Map the interactions between Sustainable Development Goals. *Nature, 534*, 320–322. https://doi.org/10.1038/534320a

Pardoe, J., Conway, D., Namaganda, E., et al. (2018). Climate change and the water–energy–food nexus: Insights from policy and practice in Tanzania. *Climate Policy, 18*, 863–877. https://doi.org/10.1080/14693062.2017.1386082

Raworth, K. (2017). The Doughnut of social and planetary boundaries (2017). In *What earth is Doughnut*? https://www.kateraworth.com/doughnut/. Accessed 2 Jan 2022.

Rockström, J., Steffen, W., Noone, K., Persson, Å., Chapin III, F. S., Lambin, E., Lenton, T. M., Scheffer, M., Folke, C., Schellnhuber, H. J., & Nykvist, B. (2009). Planetary boundaries: Exploring the safe operating space for humanity. *Ecology and Society, 14*, 1–33.

Sarkodie, S. A. (2022). Winners and losers of energy sustainability—Global assessment of the Sustainable Development Goals. *Science of the Total Environment, 154945*,. https://doi.org/10.1016/j.scitotenv.2022.154945

Spangenberg, J. H. (2002). Environmental space and the prism of sustainability: Frameworks for indicators measuring sustainable development. *Ecological Indicators, 2*, 295–309.

Steffen, W., Richardson, K., Rockström, J., et al. (2015). Planetary boundaries: Guiding human development on a changing planet. *Science, 347*, 1259855. https://doi.org/10.1126/science.1259855

UNDP. (2019). *Human Development Report 2019. Beyond income, beyond averages, beyond today: Inequalities in human development in the 21st century*. United Nations Development Programme (UNDP).

United Nations. (2015). *Transforming our world: The 2030 agenda for sustainable development*. A/RES/70/1. United Nations.

United Nations Department of Economic and Social Affairs. (2021). *The 17 goals*. Sustainable Development Goals. https://sdgs.un.org/goals. Accessed 30 Aug 2021.

West Cork Doughnut Economy Network. (2022). *Launch of the West Cork Doughnut Economy Network 24/02/2022*. https://www.youtube.com/watch?v=iEXpyoWS5CY. Accessed 2 Jan 2022.

Wu, X. F., & Chen, G. Q. (2017). Global primary energy use associated with production, consumption and international trade. *Energy Policy, 111*, 85–94. https://doi.org/10.1016/j.enpol.2017.09.024

WWF. (2018). *Living Planet Report: 2018: Aiming higher*. WWF.

CHAPTER 3

A Decarbonized Economy (Pillar Two)

Abstract This chapter presents pillar two, a decarbonized economy, focused on the immediate urgency of the climate crisis. The well-being and sustainable development trajectories of future generations critically depend on our collective efforts to realize pillar two in the present decade and to radically address the problem of global GHG emissions. Without meaningful and effective action on pillar two in the short term, our collective horizons may shift from concerns about delivering sustainable prosperity globally, to more fundamental concerns of survival for ever-increasing segments of the global population. The chapter provides an overview of the so-called 'emissions gap' between current policy pledges and emissions reductions needed in actuality to stay with the 'safe operating space' of 1.5 °C of warming, before arguing for the necessary investment to deliver a rapidly decarbonized economy, including through Green New Deal (GND) initiatives.

Keywords Decarbonized economy · GHG emissions · Sustainable development · Emissions gap · Investment · Green new deal

J. Morrissey and C. P. Heidkamp, *Demanding Sustainability*,
https://doi.org/10.1007/978-3-031-18958-6_3

'Anyone who believes in indefinite growth in anything physical, on a physically finite planet, is either mad or an economist'. Kenneth Boulding, Former Head, American Economic Association.[1]

A Frightening Emissions Gap

Launching the IPCC Sixth Assessment in August 2021, the UN Secretary-General António Guterres said *Climate Change 2021: The Physical Science Basis*, unequivocally provided 'a code red for humanity. The alarm bells are deafening, and the evidence is irrefutable' (United Nations, 2019). Boehm et al. (2021) warn that decisions made today will determine the severity of climate impacts for decades to come. Immediate, rapid and large-scale reductions in GHG emissions reductions are needed to limit warming to 1.5 °C or even 2 °C by the end of the century (UNEP, 2021a). Despite the high stakes, progress is painfully slow. For instance, G20 nations, as a group, still do not have policies in place to achieve NDCs (UNEP, 2021a). An emissions reduction pathway to limit warming to 1.5 °C would require dramatic emissions reductions, starting now and maintained over the next ten years and beyond. This transformation change is needed across all sectors: energy, food and forestry, industrial adaptation and carbon management and markets (Henderson et al., 2020). Such transformational change will need to be defined by technology transfer, capacity building and primarily, significant financial investments (Boehm et al., 2021). The IPCC is clear: 'Reducing GHG emissions across the full energy sector requires major transitions, including a substantial reduction in overall fossil fuel use, the deployment of low-emission energy sources, switching to alternative energy carriers, and energy efficiency and conservation' (IPCC, 2022, p. 36). The UNEP state that updated mitigation pledges for 2030 are to be welcomed and are beginning to move in the right direction, but the bottom line impact of these pledges on global emissions is insufficient (UNEP, 2021a). Figure 3.1 illustrates the so-called 'emissions gap', sourced from the UNEP's, 2021 Emissions Gap Report, titled 'The heat is on—A world of climate promises not yet delivered'. The figure shows that by 2030 for the

[1] Kenneth Boulding in United States Congress, House (1973), Energy reorganization act of 1973: Hearings, Ninety-Third Congress, first session, on H.R. 11510, p. 248. As cited in Kenneth McLeod, 'Learning to Think Like a Planet', Anthropocene Transitions Program Discussion Paper, October 2018, p. 5

1.5 °C goal, global annual emissions need to be 28 $GtCO_2e$ (range: 25–30 $GtCO_2e$) lower than current unconditional NDCs (UNEP, 2021a). The gap is stark, if not outright frightening in scale.

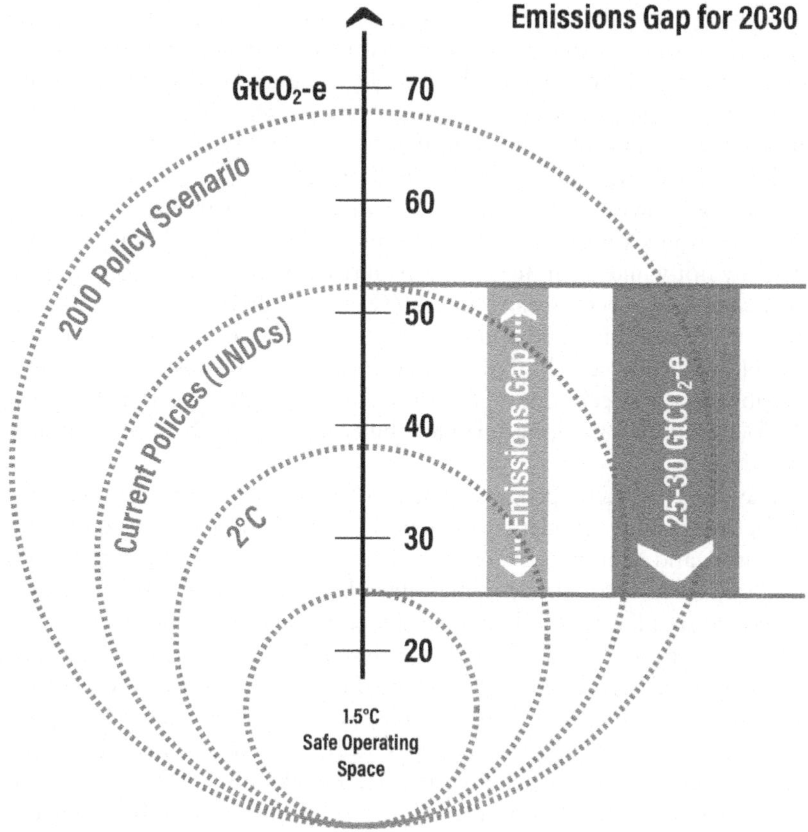

Fig. 3.1 Emissions gap for 2030[2]

[2] After the diagram presented in UNEP (2021b).

Economy as a Driver of Ecological Degradation

Despite all of this accumulating evidence, the primary economic policy objective of most governments remains fixated on the pursuit of economic growth; environmental and other policies are still evaluated first and foremost in terms of their upfront economic costs (Victor, 2012). It is now more and more obvious that the global growth-based economic system is the primary driver of global ecological degradation as captured and expressed by the planetary boundaries concept; the climate crisis is chief amongst our interlinked ecological crises but is by no means the only ecological challenge posing existential levels of threat.

There is growing recognition that a constantly expanding, extractivist-based economy cannot sustain economic growth in the long-term, and certainly not while accumulating societal risk through an erosion of essential provisioning ecological services (Özdemir, 2019, p. 626). Current levels of consumption and consequent environmental pressures are unsustainable (Marques et al., 2012). The links between well-being, welfare or happiness and economic growth are increasingly questioned (Victor, 2012), along with the long-term feasibility and sustainability of economic growth itself.

The story of social progress, whereby living conditions are improved by the increasingly efficient production of goods and services and living standards and well-being are linked to a continuous growth in the economy looks increasingly implausible (Bauhardt, 2014). While ecological economics has established a considerable body of knowledge on how to embed economies into natural ecosystems, this 'conceptual technology' has been largely ignored in mainstream economics in the post-war era. Concepts such as the *precautionary principle*[3] and the *polluter pays*[4] and have not been applied in any serious way to address the global carbon emissions challenge (Morrissey & Heidkamp, 2022). Mainstream economics has failed to consider global environmental externalities in a meaningful way, particularly when considering the temporal and spatial

[3] According to the Rio Declaration, the precautionary principle states that where there are threats of serious or irreversible damage, lack of full scientific certainty shall not be used as a reason for postponing cost-effective measures to prevent environmental degradation (United Nations, 1992).

[4] The polluter pays principle contends that firms and in turn countries should be charged for the full costs to society of their current pollution (Tilton, 2016, p. 117).

displacement of cost burdens to removed communities (e.g., Communities in the global south, communities of the future) from fossil fuel extraction and use. By and large, the global economy works on a so-called linear metabolism model. As the frontiers of new resource availability diminish and disappear, and the ecological sinks into which we deposit our wastes reach full capacity, the conceptual flaws in a linear growth-based model now increasingly manifest in real-world problems for previously removed end-consumers. While ecosystem services are estimated to be worth around US$125 trillion a year to human societies, through regulating services, water filtration, flood mitigation, crop pollination, etc., our global failure to account for such economic benefits have contributed significantly to the accumulation of ecological degradation and ecological threats (WWF, 2018). Our medium to long-term challenge is to transition to an economy based on circular metabolism, an economy which needs to be embedded within natural capital systems.

There are numerous alternative models available. O'Neill (2012) describes the *steady-state economy*: an economy where material and energy throughput is kept constant but one where throughput is also contained to within ecological limits. Van Griethuysen (2012) argues for a shift from a property-based economic rationale, towards an eco-social economic rationale, whereby social and ecological considerations and imperatives are served by economic activities, and not the other way around. The transition to a steady-state economy could be realized through *degrowth* (O'Neill, 2012), '...a voluntary transition towards a just, participatory, and ecologically sustainable society... The objectives of degrowth are to meet basic human needs and ensure a high quality of life, while reducing the ecological impact of the global economy to a sustainable level, equitably distributed between nations' (Research & Degrowth, 2010, p. 524). The degrowth concept fundamentally challenges the assumption that economic growth leads to better well-being and prosperity outcomes (Bauhardt, 2014). Van Griethuysen (2012) further outlines means to limit the pressure for economic growth by regulating capitalization and financial practices, limiting the scope of the property domain, directing investments (to 'green' initiatives) and through more efficient distribution of monetary returns.

Systemic Change

These are very radical alternatives which could not be realized without unprecedented economic (and socio-political) disruption. There are some intermediate fixes which may be applied to temper the worst aspects of ecological damage from the growth-based system, in the short term. Jackson (2017) argues for a basic macroeconomic accounting framework based on the inclusion of a wider range of indicators. This would serve to broaden economic debate beyond perennial fixations on GDP growth rates and employment levels. At the national level, the social and environmental domains of the original sustainable development concept should be afforded the same level of headline priority as the economic domain has hitherto. Such indicators are readily available to hand and have been developed based on robust research and critique. The Human Development Index (HDI) reflects key social parameters including the overall standard of living, the health of people and the level of educational attainment. The Ecological Footprint measures human demand on nature through a land-based metric. Ecological Footprint accounting considers available ecological productive land and applies this yardstick against contemporary trends of resource requirement. The associated Carbon Footprint measure has become widely popular amongst organizations and individuals. Measuring ecological impact in terms of carbon emissions in tonnes of Carbon Footprint allows consideration of individual local, regional or national impacts in the context of the overall climate crisis. A more widespread application and discussion of such metrics would emphasize the political priority afforded to ecological and social sustainability.

Wealth inequality and the externalities of consumption could be addressed through a range of taxation, pricing and incentive system reforms, which could consider the real costs which consumption patterns impose on the environment (Ripple et al., 2017). A relative decoupling view posits that a combination of a shift from goods to services, innovation and new technologies, and more extensive reuse and recycling, will successfully decouple economic growth from throughput. Such a paradigm would theoretically allow growth to continue while resource inputs and wastes decline (Victor, 2012). However, to date, empirical modelling shows no evidence for absolute decoupling occurring at a global scale, even under the most optimistic scenarios (Hickel & Kallis, 2020). Indeed, Hickel and Kallis (2020, p. 482) warn, 'that green growth remains a theoretical possibility is no reason to design policy around it when the facts are pointing in the opposite direction'. On the relationship between GDP and carbon emissions, Hickel and Kallis (2020) report that while it is still technically possible to decouple carbon emissions sufficiently to align with a global carbon budget for 1.5 or 2 °C of warming, empirical projections suggest that decoupling of this scale is unlikely.

Transformative solutions to challenges such as global energy needs will not be achieved without also transformative thinking about the structural, behavioral and political nature of these challenges (Jasanoff, 2018). Weitzman (2009) highlights the incredible magnitude of the deep structural uncertainties which climate change analysis tackles. It is essential that economics as a discipline is more explicit and open on such uncertainties. Given the scale of the challenge of carbon emissions reductions required in such a compressed timeline (over 25 gigatons reduction on annual emissions required in less than 10 years), there is a clear and overwhelming argument for large-scale mobilization of public sector resources.

While the individual may make useful contributions to sustainability through dietary shift, changes in overall consumption patterns and voting preferences at the ballot box, the infrastructure required for a low-carbon economy is large-scale, expensive and needs to be system-wide in order to have real impact. Further, stringent emissions reductions targets (determined by the science) require that the transition to a low-carbon economy happens urgently and at pace (Dolge & Blumberga, 2021).

As argued by the World Resources Institute, we need to rapidly transform our economic and societal infrastructures, 'including power generation, buildings, industry, transport, land use and agriculture—as well as the immediate scale-up of technological carbon removal' (Boehm et al., 2021, p. iv).

Green New Deal

In this regard, Green New Deal (GND) proposals argue for the coordination and financing of a large-scale transformation of the economy and of the energy system in particular (Mastini et al., 2021). GND proposals hold economic success as contingent upon an ecological restructuring of production and consumption (Bauhardt, 2014). GND concepts typically share a commitment to economic restructuring and radical transformation through adoption of Keynesian demand-side economics and large-scale investment in the public realm (Galvin & Healy, 2020). 'Green New Deal (GND) proposals are amongst the boldest initiatives for a large-scale, equality-oriented systemic transformation of Global North economies in order to address the climate crisis' (Zografos & Robbins, 2020, p. 543). GND narratives since 2018 contain features of US President Franklin D. Roosevelt's 1930's era New Deal, including social and labour reforms and public ownership of utilities (Mastini et al., 2021). GND proposals typically address long-term socio-economic prosperity in relation to employment, health and social equity in conjunction with short-term decarbonization measures, the latter a universal core feature (Boyle et al., 2021). In this way, GND concepts are informed by just transitions ideas, with consideration for both the employment implications of low-carbon transitions, as well as the vulnerabilities of communities exposed to pollution and climate impacts (White, 2020). GND approaches differ from other climate policies in two key ways: an emphasis on large-scale investment, aimed at economic transformation and a targeting of this investment to simultaneously achieve socio-economic goals, including employment, social equity and health goals (Boyle et al., 2021).

Table 3.1 presents a selection of international GND-style policy proposals, collated from the following sources: European Commission (2019a), Irfan and Golshan (2019), Macarthur et al. (2020), The United Frontline Table (UFT) (2020), Zografos and Robbins (2020), Boyle et al. (2021), and Ministry of Economy & Finance Government of South Korea (2021).

There is a wide range of diversity in the plans profilled in Table 3.1, with considerable differences in factors such as implementation, climate targets, means to address socio-economic elements and time frames considered (Boyle et al., 2021; Zografos & Robbins, 2020). However, a concern with climate action, economic opportunity and social equity is common to all (Boyle et al., 2021). As described by Allam et al. (2022) appropriately designed GND policies could accelerate sustainability transitions while pursuing social and economic justice.

Mastini et al. (2021) point to the radicalization of GND discourse when compared to earlier framings of the concept, including a rejection of the primacy of market-based environmental policy instruments and an embrace of command and control policy instruments. Schuelke-Leech (2021) underlines the transformative potential, highlighting the revolutionary implications of GND policies on lifestyles and behaviours,

Table 3.1 Summary of a selection of GND proposals

GND Proposal	*Author(s)*	*Country(ies)*	*Description*	*Goal*	*Key features*
House Resolution 109 in the US House of Representatives (2019)	Rep. Alexandria Ocasio-Cortez, 116th US congress	US	Attempt (unsuccessful) to establish a commitment by the US federal government to create a GND	Broad vision of social, economic and climate goals—core target of net zero GHG emissions by 2030	Defines 6 topical areas for action: • climate mitigation • jobs and labor • infrastructure • equity • health • education Detailed Equity and climate goals; Lacking specific detail on jobs/infrastructure
European Green Deal (EU)	European Commission	EU	Roadmap for Europe to become the first carbon–neutral continent by 2050	• No net emissions of greenhouse gases by 2050 • Economic growth decoupled from resource use • No person and no place left behind	Presently at an early implementation stage; Of the €1.8 trillion investments from the NextGenerationEU Recovery Plan and the EU's seven-year budget, one-third will finance the European Green Deal. However, NextGenerationEU includes limited specifics on environmental action beyond renewable energy development

(continued)

Table 3.1 (continued)

GND Proposal	*Author(s)*	*Country(ies)*	*Description*	*Goal*	*Key features*
Bernie Sanders' Green New Deal	Bernie Sanders, US Presidential Candidate 2020	US	Most ambitious GND plan by a US democratic candidate for the 2020 presidential election	Economy-wide and sector-specific goals with targets of 100% renewable energy for electricity and transportation by 2030 and full decarbonization of the economy by 2050	Calls for an intermediate goal of decarbonizing the two largest sources of emissions in the US transportation and energy generation, by 2030, which would lower US emissions by 71% Calls for the US to aid developing countries curb their emissions 36% by 2030
K-New Deal (July 2020–Updated July 2021)	South Korean Government	South Korea	Government program for post-Covid-19 recovery. Launched in 2020, the plan was updated in July 2021	2020 Plan—Create 1.9 million jobs by 2025 2021 Update commits to the creation of a further 600,000 jobs 2021 update contains a new category of carbon neutrality for investments to achieve the 2030 NDC	$133 billion of post-coronavirus spending targeted to digital and green sectors Original 2020 plan was not well received by environmentalists due to weak climate goals

GND Proposal	*Author(s)*	*Country(ies)*	*Description*	*Goal*	*Key features*
Pact for a Green New Deal—Canada (May 2019)	Pact for a Green New Deal—(Coalition of Civil Society Groups)	Canada	A grassroots coalition of civil society groups, unions and Indigenous peoples and scientists, launched the Pact for a Green New Deal (PGND)	100% renewable energy, oil sands phase out, 50% reduction in emissions by 2030 1 million new green jobs developed and reconciliation with Indigenous People	The mechanism to build a GND roadmap in the PGND involved more than 150 town hall meetings across Canada, with >7000 participants; Forthcoming were 'green lines' which should be included in a Canadian GND and 'red lines', which should not
A People's Orientation to a Regenerative Economy (June 2020)	The United Frontline Table (UFT) (Coalition of Civil Society Groups)	US	NGO groups including Climate Justice Alliance, It Takes Roots, and People's Action brought together 80 leaders from 64 frontline and allied organizations at the Frontline Green New Deal and Climate and Regenerative Economy Summit in Detroit (2019)	A 'Protect-Repair-Invest-Transform' Framework was applied to forward 80 priority actions, under a GND umbrella, to deliver a Regenerative Economy	Process identified green lines (what we want), yellow lines (what we're still questioning), and red lines (what we say no to) for GND policies, from development through implementation

as well as the radical innovations a GND would require across economic, social and political domains. The GND proposal forwarded in H.R. 109 argues for an interventionist economic approach to decarbonization, including a focus on industrial policies and public investments (Mastini et al., 2021). Boyle et al. (2021) argues that the breadth and ambition of GND proposals such as House Resolution 109 in the US are justified when the urgency of the climate science is considered. The European Green Deal strategy has ambitiously aimed for Europe to become the first climate-neutral continent by 2050, while in the process decoupling economic growth from resource use through a 'just and inclusive' transition (European Commission, 2019b).

Despite the transformative potential of GND approaches, GND plans have been criticized by scholars and activists as posing a potentially colonial threat, specifically to Indigenous and marginalized communities (Zografos & Robbins, 2020). Zografos and Robbins (2020, p. 543) outline the threat that 'green sacrifice zones' (GSZs) may emerge whereby the demand for specific material resources for low-carbon transitions (e.g., the use of cobalt in batteries for electric vehicles) place increased pressure on Indigenous and marginalized lands, livelihoods and sovereignties. Further, Trainer (2022, p. 1) fundamentally questions the ability of GND approaches to realize goals of decarbonization and decoupling of economic functioning and resource use, stating that 'GND goals cannot be achieved unless there is large-scale degrowth to radically different economic, social and political systems'. EEA estimates that current policy efforts to reduce GHGs, through decarbonization and improved energy efficiency, may not be adequate for realization of Green Deal 2030 targets (European Environment Agency, 2021). GND approaches, therefore, need to be considered carefully and in alignment with a diverse suite of policy instruments for just transition.

Transformative Levels of Investment

In the aftermath of the Covid-19 pandemic, calls for GND policies have become part of a recovery narrative to 'build back better' from the disruption of the pandemic and to start to meaningfully address the interlinked ecological, social and economic crises of the twenty-first century (Boyle et al., 2021). We agree with this call and further argue that an investment effort on the scale of a wartime mobilization is required to deliver a sustainable no-carbon economy, in the short time frame remaining within which the scale of the climate crisis can still be (somewhat) mitigated. As argued in Morrissey and Heidkamp (2022, p. 4), this investment should target an all of society transformation, including but not limited to: the development of public transport networks to displace reliance on private vehicle dependency; decarbonized electricity grids, developed from at-scale renewable energy investment; R&D investment in sustainable and resilient agri-food; land use reserved for biodiversity and ecosystem services, including forestry for carbon sequestration; investment in adaptation and climate resilience infrastructure, etc.

Newell and Simms (2021) point to historical precedents for such mobilizations, where states have played a central role in economic restructuring and strategization in the aftermath of significant external shocks. There is potential also for synergies and win–win outcomes, including, for example, enhancement of social protections, in conjunction with the acceleration of sustainability transitions if GND-based policies are designed appropriately (Allam et al., 2022). The post-Covid recovery period presents a possible window, therefore, where economic stimulus can be targeted to address just transitions to a no-carbon economy in a meaningful way (Morrissey & Heidkamp, 2022).

To date, developed countries have not delivered on commitments to raise US$100 billion per annum under the Paris Agreement (Iacobuta et al., 2022, p. 2). According to the World Resources Institute, climate finance remains wholly insufficient. The 'State of Climate Action 2021' report estimates that US$5 trillion is needed annually by 2030 for adequate investment levels, technology transfer and capacity building (Boehm et al., 2021). This figure broadly aligns with the US$50 trillion headline figure for capital investment to limit global warming to 1.5 °C highlighted by Wood Mackenzie (Sultoon, 2021).[5] US$2.4 trillion per annum is needed for investment in energy alone to 2035 for a 1.5 °C warming limit according to the IPCC, not considering other mitigation and adaption efforts such as land use change, coastal defenses or food production (Yeo, 2019). Project Drawdown emphasizes the upfront nature of these costs, with net operational savings accruing at a factor

[5] These investment figures comprise approximately 5.3% of global GDP (US$93.86 trillion world economy 2021, from https://statisticstimes.com).

of 4–5 times of the initial investment over the long-term (Wilkinson et al., 2020).[6] Conversely, the seminal Stem Review on the Economics of Climate Change published in 2006, warned that a failure to commit at least 1% of GDP to the climate problem in the short-medium term risked up to 20% of GDP losses in the longer term (Stern, 2007).

Box 3.1: Key Data Points—Pillar Two
US$5 trillion is needed annually by 2030 for adequate investment levels, technology transfer and capacity-building (Boehm et al., 2021).

In the immediate face of the Covid crisis, governments have rolled out measures and policy instruments which would have been unthinkable in the neo-liberal era, including nationalization of private health practices, guarantee of universal incomes and a signalled willingness to borrow extensively to fund the response (Büscher et al., 2020). The acknowledgement of the systemic nature of the financial crisis of 2008 resulted in the relatively immediate mobilization of billions for banking bailouts—calculated by Professor Deborah Lucas of MIT as amounting to a cost of US$498 billion for the US alone (Lucas, 2019).

Box 3.2: Key Data Points—Pillar Two
All available evidence suggests that the key decision-makers shaping the global economy have not accepted, or do not understand, the existential nature of the threats posed by our present ecological crises to not only the global economy but to human societies themselves.

[6] It is noteworthy that Project Drawdown suggest investment levels of $22.5–28.4 trillion, considerably lower than those indicated by either Boehm et al. (2021) or Sultoon (2021). It should be noted that the Drawdown work is framed from a bottom-up micro-analysis of solutions, as opposed to a whole-of-economy perspective and the analysis is aligned with 'Drawdown', that is, the future point in time when levels of greenhouse gases in the atmosphere stop climbing and start to steadily decline (Wilkinson et al., 2020). As such, the Drawdown analysis has not applied a hard limit of 1.5 °C by which to achieve its forwarded solutions.

Box 3.3: A decarbonized economy in action

A DECARBONIZED ECONOMY IN ACTION:

Oslo's Path to Decarbonization

Occupying 3% of the Earth's surface, cities represent nodes of strategic importance for sustainability. On the one hand, cities account for between 60 and 80% of total energy consumption and ~75% of global carbon emissions (UN, 2022). On the other hand, cities largely determine how economies grow, how resources are allocated, how innovation takes place; in addition, cities provide significant scope, efficiencies and economies of scale in carbon emissions reduction opportunities (Zenghelis & Stern, 2015).

Oslo Norway was named the European Green Capital for 2019 by the European Commission, in recognition of the city's ongoing activities and future ambitions for environmental improvement and sustainable development (World Cities Culture Forum, 2020). Oslo first set a carbon budget to drive emissions reductions in 2016, serving to 'mainstream (climate change) into daily operations and decision making'. On an annual basis, Oslo's carbon budget sets an upper limit on the total emissions of CO_2 that can be emitted by the city. The carbon budgetary constraints and parameters have been established in reference to the 1.5 °C warming scenario outlined by the Paris Agreement and a baseline of emissions related to the total emitted by the city in 2009. From the 2009 baseline, Oslo's target is to reduce city-wide GHG emissions by 95% by 2030 (Carbon Neutral Cities Alliance [CNCA], 2019). The target is therefore one of 'deep decarbonization' as defined by Linton et al. (2022).[7] Transportation, energy, buildings and waste constitute four priority sectors for decarbonization in cities. These sectors contribute the greatest proportion of city-wide emissions and also represent sectors which fall under significant municipal control/influence (Linton et al., 2022). Oslo's decarbonization/emissions reductions targets for these key sectors are reported in Linton, as follows:

Electricity	*Buildings*	*Transport*	*Waste*
Currently 100% renewable	Zero/low emissions heating in all buildings by 2020 Net-zero municipal buildings by 2020	Reduce traffic by 33% by 2030, 25% of trips by bike by 2025 100% EV by 2030	100% diversion

In addition to these targets, Oslo Pensjonsforsikring (OPF), the pension fund owned by the city, has adopted a strong strategy of divestment from fossil fuel assets based on 2 key principles: explicit consideration of climate risk in all investment decisions and a goal to reduce the carbon footprint of its equities portfolio by 40% by 2030, in line with the Paris Agreement (C40Knowledgehub.org, 2020).

Appropriate governance has been a key factor in success to date. Oslo applies a 'climate barometer' to monitor and evaluate performance of the climate budget. Furthermore, responsibility for the budget was assigned to the Vice Mayor of Finance in the city, a decision which has resulted in 'cross-municipal ownership of the climate agenda' (Nordskag, 2019).

> Look at Oslo, and you can begin to see what life will look like in a city that's serious about its obligations to the future ... In its commitment, Oslo is an outlier. Following its lead, however, cities around the world are growing interested in climate budgeting. (Romeo, 2022)

Delivering Pillar Two

Undoubtedly, pillar two represents our most acute challenge (for now). The Paris Agreement and the increasing political pressure evident at each subsequent COP, together with the emergence of more and more ambitious national targets for GHG reductions are encouraging steps in the right direction. However, progress has been characterized by a slow, if accelerating, incrementalism. The biggest challenge is time, or rather, the lack of it in which to achieve meaningful decarbonization of the economy. The following (national level) policy options might facilitate a wider transformational response to the decarbonization challenge:

- Firm commitment of up to 10% of national GDP dedicated to decarbonization in the advanced economies, in the window to 2030
- Sharp phasing out (in the short term) of subsidies to fossil fuel companies, including abolishment of all grants, tax breaks and incentives for fossil fuel exploration

[7] Linton et al. (2022, p. 2) define local deep decarbonization as a target of 80–100% net reduction in GHG emissions by 2050 or before.

- Increase in carbon taxes, with due regard for the 'low-income carbon burden'
- Increased supports for those essential industries in transition, including through regulatory change, grants, tax breaks and incentives. Examples include the agriculture sector, which needs to be rapidly directed away from the carbon-intensive production of, for example, beef and dairy, towards more ecologically regenerative production
- Substantial, incremental tax breaks and reliefs for organizations who define and publish carbon footprint analysis of their operations and commit to, and deliver on, continuous reductions on this
- Enforcement of a costed and rigorous application of the 'polluter pays' for the adverse socio-ecological impacts accruing from carbon-intensive activities, including aviation, meat consumption, land use change and fossil fuel extraction
- Regulatory transformation across government sectors to enable rapid deployment of renewable energy infrastructure, across scales from household to commercial, regional and national levels, including, for example, fast-track planning processes, incentives and feed-in-tariffs and substantial tax relief for renewable energy investment
- Direct payments to land owners for carbon sequestering ecosystem services.

References

Allam, Z., Sharifi, A., Giurco, D., & Sharpe, S. A. (2022). Green new deals could be the answer to COP26's deep decarbonisation needs. *Sustain Horizons, 1*, 100006. https://doi.org/10.1016/j.horiz.2022.100006

Bauhardt, C. (2014). Solutions to the crisis? The green new deal, degrowth, and the solidarity economy: Alternatives to the capitalist growth economy from an ecofeminist economics perspective. *Ecological Economics, 102*, 60–68. https://doi.org/10.1016/j.ecolecon.2014.03.015

Boehm, S., Lebling, K., Levin, K., et al. (2021). *State of climate action 2021.*

Boyle, A. D., Leggat, G., Morikawa, L., et al. (2021). Green new deal proposals: Comparing emerging transformational climate policies at multiple scales. *Energy Research & Social Science, 81*, 102259. https://doi.org/10.1016/j.erss.2021.102259

Büscher, B., Spierenburg, M., Bavinck, M., et al. (2020). *Planning for post-corona: Five proposals to craft a radically more sustainable and equal world.*

C40Knowledgehub.org. (2020). Oslo's journey to decarbonisation. In *Case Studies Best Practice*. Examples. https://www.c40knowledgehub.org/s/article/Oslos-journey-to-decarbonisation?language=en_US. Accessed 5 Jan 2022.

Carbon Neutral Cities Alliance (CNCA). (2019). Set a climate budget to drive deep decarbonization. In *How City Oslo Management Carbon Like It Management Money*. https://carbonneutralcities.org/set-a-climate-budget-to-drive-deep-decarbonization/. Accessed 2 Jan 2022.

Dolge, K., & Blumberga, D. (2021). Economic growth in contrast to GHG emission reduction measures in Green Deal context. *Ecological Indicators, 130*. https://doi.org/10.1016/j.ecolind.2021.108153

European Commission. (2019a). A European green deal. In *Priorities 2019a–2024*. https://ec.europa.eu/info/strategy/priorities-2019a-2024/european-green-deal_enface. Accessed 10 Jun 2022.

European Commission. (2019b). Communication from the commission to The European Parliament, The European Council, The Council, The European Economic and Social Committee and The Committee of the Regions—The European Green Deal. Brussels.

European Environment Agency. (2021). *Trends and projections in Europe 2021*.

Galvin, R., & Healy, N. (2020). The green new deal in the United States: What it is and how to pay for it. *Energy Research Social Science, 67*. https://doi.org/10.1016/j.erss.2020.101529

Henderson, K., Pinner, D., Rogers, M., et al. (2020). *Climate math: What a 1.5-degree pathway would take*. McKinsey Q 1–18.

Hickel, J., & Kallis, G. (2020). Is green growth possible? *New Political Economy, 25*, 469–486.

Iacobuta, G. I., Brandi, C., Dzebo, A., et al. (2022). Aligning climate and sustainable development finance through an SDG lens: The role of development assistance in implementing the Paris Agreement. *Global Environmental Change, 74*, e102509. https://doi.org/10.1016/j.gloenvcha.2022.102509

IPCC. (2022). Summary for policymakers makers. In P. R. Shukla, J. Skea, & R. Slade et al. (Eds.), *Climate change 2022: Mitigation of climate change. Contribution of working group III to the sixth assessment report of the intergovernmental panel on climate change* (pp. 1–64). Cambridge University Press.

Irfan, U., & Golshan, T. (2019). Bernie Sanders's Green New Deal, explained. In *Vox's Guide to Where 2020 Democracy Stand Policy*. https://www.vox.com/2019/8/22/20827396/bernie-sanders-2020-climate-policy-green-new-deal. Accessed 10 Jun 2022.

Jackson, T. (2017). *Prosperity without growth: Foundations for the economy of tomorrow* (2nd ed.). Routledge.

Jasanoff, S. (2018). Just transitions: A humble approach to global energy futures. *Energy Research & Social Science, 35*, 11–14. https://doi.org/10.1016/j.erss.2017.11.025

Linton, S., Clarke, A., & Tozer, L. (2022). Technical pathways to deep decarbonization in cities: Eight best practice case studies of transformational climate mitigation. *Energy Research & Social Science, 86*, 102422. https://doi.org/10.1016/j.erss.2021.102422

Lucas, D. (2019). Measuring the cost of bailouts. *Annual Review of Financial Economics, 11*, 85–108. https://doi.org/10.1146/annurev-financial-110217-022532

Macarthur, J. L., Hoicka, C. E., Castleden, H., et al. (2020). Canada's Green New Deal: Forging the socio-political foundations of climate resilient infrastructure? *Energy Research & Social Science, 65*, 101442. https://doi.org/10.1016/j.erss.2020.101442

Marques, A., Rodrigues, J., Lenzen, M., & Domingos, T. (2012). Income-based environmental responsibility direct. *Ecological Economics, 84*, 57–65. https://doi.org/10.1016/j.ecolecon.2012.09.010

Mastini, R., Kallis, G., & Hickel, J. (2021). A green new deal without growth? *Ecological Economics, 179*, 106832. https://doi.org/10.1016/j.ecolecon.2020.106832

Ministry of Economy & Finance Government of South Korea. (2021). *Government announces Korean new deal 2.0.*

Morrissey, J., & Heidkamp, C. P. (2022). Sustainability after Covid: Pillars for a just transition. *Environmental Sustainability*. https://doi.org/10.1007/s42398-022-00231-y

Newell, P., & Simms, A. (2021). How did we do that? Histories and political economies of rapid and just transitions. *New Political Economy, 26*, 907–922. https://doi.org/10.1080/13563467.2020.1810216

Nordskag, M. (2019). *How Oslo manages carbon emissions like it manages money*. Fast Co. https://www.fastcompany.com/90410124/how-oslo-manages-carbon-emissions-like-it-manages-money. Accessed 2 Jan 2022.

O'Neill, D. (2012). Measuring progress in the degrowth transition to a steady state economy. *Ecological Economics, 84*, 221–231. https://doi.org/10.1016/j.ecolecon.2011.05.020

Özdemir, V. (2019). Innovating governance for planetary health with three critically informed frames. *OMICS: A Journal of Integrative Biology, 23*, 623–630. https://doi.org/10.1089/omi.2019.0175

Research & Degrowth. (2010). Degrowth declaration of the Paris 2008 conference. *Journal of Cleaner Production, 18*, 523–524.

Ripple, W. J., Wolf, C., Newsome, T. M., et al. (2017). World scientists' warning to humanity: A second notice. *BioScience, 67*, 1026–1028.

Romeo, N. (2022). *How Oslo learned to fight climate change.*

Schuelke-Leech, B. A. (2021). Disruptive technologies for a green new deal. *Current Opinion in Environmental Science and Health, 21*, 100245. https://doi.org/10.1016/j.coesh.2021.100245

Stern, N. (2007). *Stern review: The economics of climate change*. Cambridge University Press.

Sultoon, J. (2021). What would it take to limit global warming to 1.5 degrees? In Energy Transit. https://www.woodmac.com/news/opinion/what-would-it-take-to-limit-global-warming-to-1.5-degrees/. Accessed 10 Jun 2022.

The United Frontline Table (UFT). (2020). *A people's orientation to a regenerative economy*.

Tilton, J. E. (2016). Global climate policy and the polluter pays principle: A different perspective. *Resources Policy, 50*, 117–118. https://doi.org/10.1016/j.resourpol.2016.08.010

Trainer, T. (2022). A technical critique of the Green New Deal. *Ecological Economics, 195*, 107378. https://doi.org/10.1016/j.ecolecon.2022.107378

UN. (2022). Goal 11: Make cities inclusive, safe, resilient and sustainable. In *Sustainable Development Goals*. https://www.un.org/sustainabledevelopment/cities/. Accessed 4 May 2022.

UNEP. (2021a). *Emissions gap report 2021a: The heat is on—A world of climate promises not yet delivered—Executive Summary*. Nairobi.

UNEP. (2021b). *Last call for climate action*. https://wedocs.unep.org/bitstream/handle/20.500.11822/37341/LCCA_EGI.pdf. Accessed 2 Jan 2022.

United Nations. (1992). Rio declaration on environment and development. In *Report of the United Nations conference on environment and development* (pp. 3–14).

United Nations. (2019). IPCC report: 'Code red' for human driven global heating, warns UN chief. In *UN News Global Perspective Human Stories*. https://news.un.org/en/story/2021/08/1097362#:~:text=%E2%80%99Coderedforhumanity%E2%80%99&text=Weareatimminentrisk,%2Ctokeep1.5alive.%22. Accessed 12 Jun 2022.

Van Griethuysen, P. (2012). Bona diagnosis, bona curatio: How property economics clarifies the degrowth debate. *Ecological Economics, 84*, 262–269. https://doi.org/10.1016/j.ecolecon.2012.02.018

Victor, P. A. (2012). Growth, degrowth and climate change: A scenario analysis. *Ecological Economics, 84*, 206–212. https://doi.org/10.1016/j.ecolecon.2011.04.013

Weitzman, M. L. (2009). On modeling and interpreting the economics of catastrophic climate change. *Review of Economics and Statistics, 91*, 1–19.

White, D. (2020). Just transitions/design for transitions: Preliminary notes on a design politics for a green new deal. *Capital Nature, Social, 31*, 20–39. https://doi.org/10.1080/10455752.2019.1583762

Wilkinson, K., Chissell, C., & Foley, J. 2020. *The drawdown review: Climate solutions for a new decade.*

World Cities Culture Forum. (2020). *Creative climate cities profile Oslo.*

WWF. (2018). *Living planet report: 2018: Aiming higher*. WWF, Gland, Switzerland.

Yeo, S. (2019). Climate finance: The money trail. *Nature, 573*, 328–331.

Zenghelis, D., & Stern, N. (2015). *Climate change and cities: A prime source of problems, yet key to a solution.* Guard.

Zografos, C., & Robbins, P. (2020). Green sacrifice zones, or why a green new deal cannot ignore the cost shifts of just transitions. *One Earth, 3*, 543–546. https://doi.org/10.1016/j.oneear.2020.10.012

CHAPTER 4

A Shared (Cost) Burden (Pillar Three)

Abstract Profound distributive injustices underly global patterns of resource use and related carbon emissions. In addition, those with the least responsibility for the climate crisis are projected to be most severely impacted in the decades ahead, while those with the greatest share of resources continue to produce disproportionate levels of pollution. Those with the resources need to do much more, particularly as countries facing the highest levels of temperature variability also have the least economic potential to cope with the impacts. A fair sharing of climate change burdens between countries would mean that those countries benefiting from greater climate stability would also bear a higher burden of climate change mitigation and adaptation efforts.

Keywords Distributive injustice · Pollution · Fair share · Climate burden · Mitigation · Adaptation

> *Taking future generations into account in a fair distribution of resources and other means of development should still be strived for as a long-term objective; requiring a cultural maturity that is a far cry from the current individualist and consumerist propensity.* (Van Griethuysen, 2012, p. 266)

J. Morrissey and C. P. Heidkamp, *Demanding Sustainability*,
https://doi.org/10.1007/978-3-031-18958-6_4

Carbon Injustice

The seminal Stern Review concluded that climate change is 'the greatest market failure the world has ever seen'; that all countries will be impacted by climate change, with the poorest countries likely to suffer earliest and most severely; Stern also warned that inaction was likely to be far more expensive in the longer term that 'the costs of action' (Stern, 2007). Stern flagged issues of cost–benefit asymmetry and distributional justice:

> The costs of taking action are not evenly distributed across sectors or around the world... developing countries should not be required to bear the full costs of this action alone. (Stern, 2007, p. vii)

The unsustainability and inherent inequality of global consumption patterns and carbon-intensive lifestyles (for some) are obvious, if still in need of urgent debate and critical discussion (Hubacek et al., 2017). Oxfam point to the inextricable links between climate change and economic inequality—the climate crisis driven by the emissions of the 'haves' while the 'have-nots' bear the worst of adverse consequences (Gore, 2015). The richest 1% produce twice the total amount of carbon emitted by the poorest 50% of humanity (Oxfam, 2020). The G20 nations account for 78% of all emissions (UNEP, 2019). The average carbon footprint of the globe's richest 1% may dwarf the footprint of the poorest 10% by as much as a factor of 175 (Gore, 2015). To compound matters, the richest 1% are projected to contribute an even bigger and more disproportionate share of total global emissions by 2030, compared with 2015 when the Paris Agreement was signed (Gore, 2021).

Developing countries in the Global South will be disproportionally impacted due to factors such as low latitude location, lagging development levels and the amount and variability of rainfall (Stern, 2007). The lowest per capita CO_2 emissions are observed for countries in sub-Saharan Africa. Simultaneously, these countries are ranked as the most vulnerable to adverse climate change impacts by institutions including the World Bank and United Nations (Eckstein et al., 2019). African countries, in particular, are highly vulnerable to climate change impacts because of the combination of low adaptive capacity with factors of space and place, including particular geographical, environmental and socio-economic conditions (Eckstein et al., 2019). The IPCC highlight that considerable proportions of the populations of low emitting countries lack access to modern energy services (IPCC, 2022, p. 9), where issues of energy access and energy poverty represent a significant barrier to SDG realization. Climate change is an ever increasing and acute threat to global development and efforts to end poverty. Eckstein et al. (2019) warn that an additional 100 million people could be pushed into poverty by climate impacts by 2030 (Eckstein et al., 2019).

Unsustainable Cost Burdens

Swiss Re report a projected loss in global GDP of up to 10% if Paris agreement commitments are not delivered on (Swiss Re Institute, 2021). Under a 1.5 °C warming scenario, accumulated costs from climate-related damage may be as high as US$54 trillion by 2100,[1] rising to US$69 trillion under a 2 °C warming scenario (Hoegh-Guldberg et al., 2018). These costs will certainly fall disproportionately on communities in developing countries (Black, 2022). Swiss Re projects losses of 33–36% of GDP by 2048 for Thailand, Malaysia and the Philippines under scenarios of 2–2.6 °C of warming, for instance, with many other countries in South and Southeast Asia identified as being particularly vulnerable to adverse climate-related impacts (Swiss Re Institute, 2021). In addition to South Asia, Kikstra et al. (2021) projects that climate change will cause deteriorating growth outlooks for Africa and Latin America. Countries of the Global North, by contrast, characterized by stronger capacities to respond to shocks as well as lower exposure levels, are considerably less vulnerable (Swiss Re Institute, 2021). Further, it is also likely that the cost of government borrowing, all else being equal, will be impacted and influenced by vulnerability and resilience to climate change (Cevik & Jalles, 2022), compounding existing inequalities.

Box 4.1: Key Data Points—Pillar Three
78% of all global emissions are produced by the G20 group of nations (UNEP, 2019).

Climate change is thus unfolding in a context of disparate historical development pathways, including the disparate use of, and benefit

[1] Relative to 1961–1990.

from, historical carbon usage (Morrissey & Heidkamp, 2022). Teng et al. (2011) use the concept of 'carbon space' to frame their analysis of global inequality in the era of climate change. Considering the historical cumulative emission of major countries from 1850 to 2006, Teng et al. conclude that 70% of 'carbon space' has been distributed unequally. The UNEP estimate that the carbon budget to limit warming to 1.5 °C relative to pre-industrial levels, that is, the amount of carbon which humanity can collectively emit and still stay within 1.5 °C of warming, is approximately 400 $GtCO_2$. This carbon budget figure rises to 1150 $GtCO_2$ for a 2 °C warming scenario (UNEP, 2021). Alcaraz et al. (2018) outline a method by which this global carbon budget might be distributed in a fair manner, using climate justice criteria. The Model of Climate Justice per capita (MCJ) posits that those countries characterized by high population levels and low historical carbon emissions should be allocated higher carbon budgets, under sustainability and justice criteria. Such countries include Brazil, Other Latin American and Caribbean (OLAC) countries, India and Other African (OA) countries—all characterized by real development needs and by low historical emissions. Considerably reduced carbon budgets should subsequently be allocated for the EU27 and the US, i.e., advanced, wealthy countries with high historical emission levels (when compared globally on a per country basis). Morrissey and Heidkamp (2022) applied Alcaraz's MCJ method to the UNEP carbon budget figure of 1150 $GtCO_2$ for 2 °C of warming. Using this approach, the following carbon budget emissions figures to 2100 were calculated: USA-31Gt, EU27-63Gt, Brazil-33Gt, India-220Gt, Other African-219Gt. The figures point to a challenge of enormous scale, requiring unprecedented mobilization of economic, social and political capital. Clearly, there is little evidence of such a historical mobilization at present (Morrissey & Heidkamp, 2022).[2]

The non-reversibility of climate changes is important. Climate change induced sea-level rise is a millennia-long process, which in human time frames is essentially non-reversible. Small island developing states such as Tuvalu and Kiribati in the Pacific are highly vulnerable to climate impacts (Cauchi et al., 2020); these states represent a 'canary in the coal mine' warning to the global community of the potential devastating impacts of climate change. A family of four from Tuvalu was granted residency in

[2] G20 members are not collectively on track to achieve 2030 emissions reduction pledges (UNEP, 2021).

New Zealand in 2014, widely heralded as the first legal recognition of 'climate change refugees'.[3] On Tuvalu and on Kiribati rising tides have resulted in salt water encroachment which has impacted on soil viability, has resulted in the salinization of drinking water supplies and which has isolated damaged or undermined key infrastructure, including transport housing and community infrastructure. Environmentally driven migration is already taking place around and out of Tuvalu (Milan et al., 2016). The impacts of climate change on these communities are present now and are of a severe nature. However, the contributions on a per capita basis from low-lying Pacific islands such as Kiribati and Tuvalu are minuscule when compared to per capita emissions from the Global North. CO_2 emissions per capita for Kiribati stood at 0.6 tonnes of CO_2 in 2018, for instance. This compares with an EU-27 average of 6.7 tonnes of CO_2 per capita for the same period (Our World in Data, 2021).

Box 4.2: Key Data Points—Pillar Three
CO_2 emissions for Kiribati a vulnerable, low-lying nation in the Pacific, are 0.6 tonnes of CO_2 per capita, in stark contrast to the EU-27 average of 6.7 tonnes of CO_2 per capita (Our World in Data, 2021).[4]

Figure 4.1 shows the distributive injustice in the cause-effect dynamics of the climate crisis. Based on data and analysis by Oxfam and the Institute for European Environmental Policy (Gore, 2021; Oxfam, 2020). Figure 4.1 illustrates that the emissions of the richest 10% were comparable with the rest of the world combined, over the time-period 1990–2015. Further, the richest 1% were responsible for double the carbon emissions of the poorest 50% over this time. These historical injustices are also projected to persist and worsen. By 2030, the emissions of the richest 1% are set to exceed the 1.5 °C-compatible per capita level by a factor of 30. At the same time, the poorest 50% will remain below 1.5 °C-compatible per capita levels of emissions.

[3] McAdam argues that the decision was based purely on humanitarian and discretionary grounds. Nevertheless, New Zealand has started to systematically delineate the legal protection framework applicable to claims based on the impacts of climate change, natural disasters or environmental degradation (McAdam, 2015).

[4] Data for 2018.

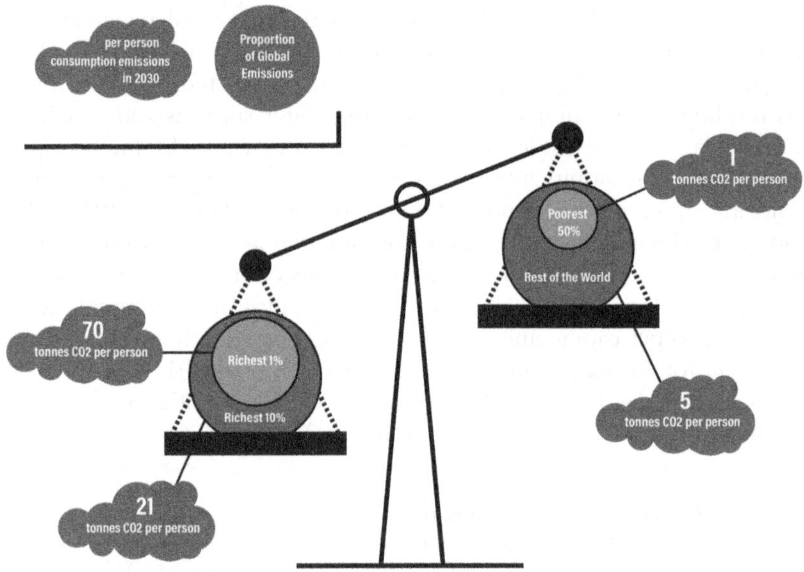

Fig. 4.1 Climate cause and effect—an unequal load[5]

Paying up for Climate Action

Achieving the Paris climate targets would require rapid, aggressive and committed decarbonization in rich countries but could legitimately limit growth aspirations of poor countries, given the current fossil-dependent growth models (Hubacek et al., 2017; Steckel et al., 2013). The study by Campagnolo and Davide (2019) shows that a full implementation of emission reduction contributions (as per NDCs of the Paris Agreement), would impede efforts to reduce poverty by 2030 (+4.2% of the population below the poverty line compared to the baseline scenario). However, if appropriate financial support for climate mitigation in developing countries were to be provided through international climate funding mechanisms, aggregate poverty levels could be slightly reduced (Campagnolo & Davide, 2019). The key implication is that despite the high costs associated with the implementation of mitigation actions, countries in the Global South face the threat of even higher costs

[5] After Oxfam (2020) and Gore (2021).

potential to cope with adverse climate impacts (Bassetti, 2019). In addition, unsustainable debt burdens make financing for climate adaptation more difficult and reduce capacity to recover from loss and damage after climate-related disruptions (Fresnillo, 2020). Fresnillo (2020, p. 3) provides clear guidance on how to address the linked impacts of sovereign debt and climate crises:

- Review debt sustainability
- Timely and sufficient debt relief
- Comply with climate finance commitments
- Provide finance to address loss and damage
- Provide emergency additional finance
- Debt payments suspension and debt relief in case of climate disaster
- Debt workout mechanism.

These 7 action points further highlight the need for enhanced international effort on resilience building, adaptation and mitigation of 'climate debt trap' risks (Bassetti, 2019); the importance of meaningful progress on the SDGs is further underscored (Running, 2015).

Box 4.3: Key Data Points—Pillar Three

> ...a massive transformation is needed to unlock the trillions required to help the world shift to a low-carbon future and build resilience to climate change. Financiers will have to step away from approaching climate change on a project-by-project basis—a wind farm here, a solar plant there—and start thinking about the carbon impact of every dollar spent. (Yeo, 2019)

Box 4.4: A Shared (Cost) Burden in Action
A SHARED (COST) BURDEN IN ACTION:
Water Sector Resilience Nexus for Sustainability in Barbados (WSRN S-Barbados)

Barbados is a water-stressed country, with the current level of water availability and access insufficient to satisfy overall demand. Climate change is already having a serious impact on the country, including impacts from hurricanes and tropical storms. Water availability is one impact expected if inaction persists (Antimiani et al., 2017). Even slight variatioı climatic conditions can precipitate large costs in developing coun For instance, hydrological variability may reduce average annual (growth rates in Ethiopia by as much as 38% and increase po\ rates by up to 25% (Stern, 2007; World Bank, 2006). The evid is mounting that climate disasters, and associated further indebted occurring subsequently, are creating serious challenges for developr (Bassetti, 2019).

Developed nations have made firm commitments under The I Agreement to make increased financial flows available to developing cc tries, starting from the 2020 pledge to mobilize US$100 billion year (Campagnolo & Davide, 2019). The Green Climate Fund (G is the primary means to provide compensation under the current clin regime, with the key purpose of providing support to developing co tries to limit or reduce GHG emissions and to help with adaptat efforts (Antimiani et al., 2017). Under the UN Framework Convent on Climate Change, NDCs outlined in the Paris Agreement are m on a voluntary basis, rights and duties are not made explicit. Simil for SDGs, whereby national targets in line with the 2030 Agenda determined by the countries themselves (Häyhä et al., 2016).

Equity-focused approaches allocate emissions according to histor or current responsibilities, framed in terms of (cumulative) GHG po tion emitted to date to the global atmosphere (Hubacek et al., 201 Fleurbaey et al. (2014) outlines four key equity principles: capac responsibility, equality and the right to sustainable development. A rec tribution of carbon shares from global elites to the global poor wo seem to be compatible with distributive justice principles, particula given international poverty alleviation goals (Hubacek et al., 2017). Su an approach would inevitably involve a reduction of wealth inequal globally and importantly, reduced per capita consumption in the advanc economies of the Global North (Ripple et al., 2017). Fair climate chan burden sharing between countries should ensure that those countr benefiting from greater climate stability could and probably should al bear a higher burden of the costs of climate change mitigation and ada tation (Puaschunder, 2020). Such burden-sharing would address the cos of responding to climate change, as well as issues of unsustainable de burdens. For vulnerable developing countries, environmental hazards ca increase the cost of borrowing as well as the risk of debt crises; mar such vulnerable countries are already highly indebted to begin (Fre nillo, 2020). Many highly vulnerable countries also have low econom to exacerbate in the coming years and decades. At present, households experience water insecurity, typified by frequent outages in some parts of the country that can last for weeks on end (Wilson, 2022). The Green Climate Fund (GCF), the Caribbean Community Climate Change Centre (CCCCC) and the Barbados Water Authority (BWA) are cooperating to transform the water sector in Barbados (Green Climate Fund, 2021). In 2018, the GCF contributed US$27 million for the Water Sector Resilience Nexus for Sustainability project, as part of an overall US$45.2 million package. The project began in Jan 2019 and is expected to have a lifespan of 30 years, with 220.2 k projected tonnes of emissions avoided and over 470,000 beneficiaries of the scheme. The overall aims of the GCF project in Barbados are to (Green Climate Fund, 2022):

- Improve access to potable water
- Increase resilience to extreme climate events
- Reduce water disruptions.

WSRN S-Barbados aims to increase water security through significant infrastructure development, including the installation of solar photovoltaic (PV) and natural gas backup power for pumping stations (Wilson, 2022). The installation of solar panels in water pumping stations will provide clean energy, cut costs and stabilize access to electricity. Saved money can then be reinvested in water infrastructure upgrades, including upgrading the mains supply (Green Climate Fund, 2021). The project also involves funding an initiative of potable water tanks across the country. These tanks provide a local store of water and are equipped with pumping technology to supply households, schools or businesses with water if the mains system cannot do so. Approximately 1500 such tanks are targeted under WSRN S-Barbados (Wilson, 2022).

The project is the first GCF project to be implemented in the Caribbean and is expected to provide a replicable framework for countries in the Eastern Caribbean experiencing similar water–energy–climate challenges (Green Climate Fund, 2021). While on the one hand, projects like this one point the potential of the GCF to catalyze measurable and meaningful change in addressing the climate crisis (Puri et al., 2022), current levels of investment and finance provision more broadly are wholly inadequate when compared to the scale of the challenge at hand, e.g., about US$900 billion should be invested annually up to 2030 to meet Paris targets in the energy sector alone (Kiremu et al., 2022). The report by Binet et al. (2021) provides a useful critical overview of the approach of the GCF.

Delivering Pillar Three

While each pillar of this volume presents considerable political challenges, pillar three *a shared (cost) burden* is arguably the most challenging of all, straddling as it does, the deepest fault line historically in the left–right political divide. The pillar is characterized by both normative and practical dimensions. We argue that wealth redistribution ***should*** happen from a justice point of view because it is ethically the right thing to do. However, from a practical perspective, wealth redistribution ***must*** happen if we are to have any chance of addressing our linked economic, social and ecological crises. Fresnillo's (2020) excellent analysis provides key recommendations which we reiterate here as key means to deliver on pillar three. These are augmented with some additional observations of our own.

- The Paris Agreement has identified increased financial flows to developing countries as a key mechanism to deliver on global mitigation. The absolute amounts to be transferred need to be drastically increased, in the short term, in line with a 1.5 °C warming pathway
- Further, the rights and duties of advanced economies in terms of climate finance commitments, loss and damage and emergency additional finance need to be made more explicit and a binding framework to honor commitments established
- Fresnillo's recommendations of debt sustainability reviews and relief, appropriate finance for emergency response and loss and damage and debt relief in case of climate disaster are important and likely to become even more significant in the coming years
- As with other pillars, pillar three requires an extensive national dialogue on 'the right thing to do' and a reevaluation of national interests and priorities, particularly in the advanced economies. The cost–benefit dynamics of large-scale wealth transfers to the Global South need to be interrogated and more widely understood, specifically with regard to the longer-term benefits to overall prosperity for everyone from short-term carbon mitigation financing for those countries most in need of assistance
- Wide-spread societal buy-in in the advanced economies could identify their 'carbon debt' to the Global South through an analysis and identification of historical cumulative carbon emissions and an acknowledgement of the 'carbon space' which their nation has 'occupied'

- Personal taxation related to carbon emissions beyond a threshold amount would enable explicit acknowledgment of the global impact of individual actions and would target high-impact lifestyles. This would need to be carefully applied in the context of low-income carbon burdens
- Global wealth inequality has become preposterous in the aftermath of the COVID-19 pandemic. Oxfam (2022) reports that 252 men have more wealth than the entire 1 billion population of women and girls in Latin America, the Caribbean and Africa combined. Humanity cannot afford to support this billionaire class and their disproportionate and reckless environmental impacts. Extreme wealth taxes are essential.

References

Alcaraz, O., Buenestado, P., Escribano, B., et al. (2018). Distributing the global carbon budget with climate justice criteria. *Climate Change, 149*, 131–145.

Antimiani, A., Costantini, V., Markandya, A., et al. (2017). The Green Climate Fund as an effective compensatory mechanism in global climate negotiations. *Environmental Science & Policy, 77*, 49–68. https://doi.org/10.1016/j.envsci.2017.07.015

Bassetti, F. (2019). The cruel irony of climate debt. In *Migrations and inequalities, Foresight: The CMCC observatory on climate policies and futures*. https://www.climateforesight.eu/migrations-inequalities/the-cruel-irony-of-climate-debt/. Accessed 19 Jan 2021.

Binet, S., De Bruijn, M., Horikoshi, D., et al. (2021). *Independent evaluation of the adaptation portfolio and approach of the Green Climate Fund*. Evaluation. Independent Evaluation Unit, Green Climate Fund.

Black, R. (2022). *Climate economics—Costs and benefits*. Energy and Climate Intelligence Unit. Briefings. https://eciu.net/analysis/briefings/climate-impacts/climate-economics-costs-and-benefits#:~:text=Thecostsofclimatechange-The impacts of&text=TheUNEnvironmentProgrammeestimated,billionperyearby2050. Accessed 14 Feb 2022.

Campagnolo, L., & Davide, M. (2019). Can the Paris deal boost SDGs achievement? An assessment of climate mitigation co-benefits or side-effects on poverty and inequality. *World Development, 122*, 96–109. https://doi.org/10.1016/j.worlddev.2019.05.015

Cauchi, J. P., Moncada, S., Bambrick, H., & Correa-velez, I. (2020). Coping with environmental hazards and shocks in Kiribati: Experiences of climate change by atoll communities in the Equatorial Pacific. *Environmental Development, 37*, 100549. https://doi.org/10.1016/j.envdev.2020.100549

Cevik, S., & Jalles, J. T. (2022). This changes everything: Climate shocks and sovereign bonds. *Energy Economics, 107*, 105856. https://doi.org/10.1016/j.eneco.2022.105856

Eckstein, D., Hutfils, M., & Winges, M. (2019). *Global Climate Risk Index 2019. Who suffers most from extreme weather events? Weather-related loss events in 2017 and 1998 to 2017*. Germanwatch e.V.

Fleurbaey, M., Kartha, S., Bolwig, S., et al. (2014). Sustainable development and equity. In O. Edenhofer, R. Pichs-Madruga, Y. Sokona, et al. (Eds.), *Climate change 2014: Mitigation of climate change. Working group III contribution to the fifth assessment report of the Intergovernmental Panel on Climate Change* (pp. 238–350). IPCC.

Fresnillo, I. (2020). *A tale of two emergencies: The interplay of sovereign debt and climate crises in the global south*. Eurodad.

Gore, T. (2015). *Extreme carbon inequality: Why the Paris climate deal must put the poorest, lowest emitting and most vulnerable people first*. Oxfam.

Gore, T. (2021). *Carbon inequality in 2030*. Oxford.

Green Climate Fund. (2021). *GCF in Barbados: Creating a clean energy, climate-resilient water sector*. https://www.youtube.com/watch?v=MSnKQM6f5ow. Accessed 2 Feb 2022.

Green Climate Fund. (2022). *FP060 water sector resilience nexus for sustainability in Barbados (WSRN S-Barbados)*. https://www.greenclimate.fund/project/fp060. Accessed 9 May 2022.

Häyhä, T., Lucas, P. L., van Vuuren, D. P., et al. (2016). From Planetary Boundaries to national fair shares of the global safe operating space—How can the scales be bridged? *Global Environmental Change, 40*, 60–72. https://doi.org/10.1016/j.gloenvcha.2016.06.008

Hoegh-Guldberg, O., Jacob, D., Taylor, M., et al. (2018). Impacts of 1.5°C of global warming on natural and human systems. In V. Masson-Delmotte, P. Zhai, H.-O. Pörtner, et al. (Eds.), *Global warming of 1.5°C. An IPCC Special Report on the impacts of warming of 1.5°C above pre-industrial levels and related global greenhouse gas emission pathways, in the context of strengthening the global response to the threat of climate change, sustainable development, and efforts to eradicate poverty*. The Intergovernmental Panel on Climate Change (IPCC).

Hubacek, K., Baiocchi, G., Feng, K., et al. (2017). Global carbon inequality. *Energy, Ecology and Environment, 2*, 361–369. https://doi.org/10.1007/s40974-017-0072-9

IPCC. (2022). Summary for policymakers makers. In P. R. Shukla, J. Skea, R. Slade, et al. (Eds.), *Climate Change 2022: Mitigation of climate change*.

Contribution of Working Group III to the Sixth Assessment Report of the Intergovernmental Panel on Climate Change (pp. 1–64). Cambridge University Press.

Kikstra, J. S., Waidelich, P., Rising, J., et al. (2021). The social cost of carbon dioxide under climate-economy feedbacks and temperature variability. *Environmental Research Letters, 16*, 094037.

Kiremu, M., Scrimgeour, F., Mutegi, J., et al. (2022). Climate finance readiness: A review of institutional frameworks and policies in Kenya. *Sustainable Environment, 8*. https://doi.org/10.1080/27658511.2021.2022569

McAdam, J. (2015). The emerging New Zealand jurisprudence on climate change, disasters and displacement. *Migration Studies, 3*(1), 131–142.

Milan, A., Oakes, R., & Campbell, J. (2016). *Tuvalu: Climate change and migration relationships between household vulnerability, human mobility and climate change report*. United Nations University Institute for Environment and Human Security.

Morrissey, J., & Heidkamp, C. P. (2022). Sustainability after COVID: Pillars for a just transition. *Environmental Sustainability*. https://doi.org/10.1007/s42398-022-00231-y

Our World in Data. (2021). *CO_2 and greenhouse gas emissions*. https://ourworldindata.org/co2-and-other-greenhouse-gas-emissions. Accessed 19 Jan 2021.

Oxfam. (2020). *The carbon inequality era*. https://oxfam.medium.com/the-carbon-inequality-era-71e20205a0d4. Accessed 2 Feb 2022.

Oxfam. (2022). A deadly virus: 5 shocking facts about global extreme inequality. https://www.oxfam.org/en/5-shocking-facts-about-extreme-global-inequality-and-how-even-it. Accessed 10 Aug 2022.

Puaschunder, J. (2020). *Mapping climate justice*. Palgrave Macmillan.

Puri, J., Prowse, M., Roy, E. D., & Huang, D. (2022). Assessing the likelihood for transformational change at the Green Climate Fund: An analysis using self-reported project data. *Climate Risk Management, 35*, 100398. https://doi.org/10.1016/j.crm.2022.100398

Ripple, W. J., Wolf, C., Newsome, T. M., et al. (2017). World scientists' warning to humanity: A second notice. *BioScience, 67*, 1026–1028.

Running, K. (2015). Towards climate justice: How do the most vulnerable weigh environment—Economy trade-offs? *Social Science Research, 50*, 217–228. https://doi.org/10.1016/j.ssresearch.2014.11.018

Steckel, J. C., Brecha, R. J., Jakob, M., et al. (2013). Development without energy? Assessing future scenarios of energy consumption in developing countries. *Ecological Economics, 90*, 53–67. https://doi.org/10.1016/j.ecolecon.2013.02.006

Stern, N. (2007). *Stern review: The economics of climate change*. Cambridge University Press.

Swiss Re Institute. (2021). *The economics of climate change: No action not an option*. Swiss Re Institute.

Teng, F., He, J., Pan, X., & Zhang, C. (2011). Metric of carbon equity: Carbon Gini Index based on historical cumulative emission per capita. *Advances in Climate Change Research, 2*, 134–140. https://doi.org/10.3724/SP.J.1248.2011.00134

UNEP. (2019). *Emissions Gap Report 2019*. United Nations Environment Programme.

UNEP. (2021). *Emissions Gap Report 2021: The heat is on—A world of climate promises not yet delivered—Executive Summary*. UNEP.

Van Griethuysen, P. (2012). Bona diagnosis, bona curatio: How property economics clarifies the degrowth debate. *Ecological Economics, 84*, 262–269. https://doi.org/10.1016/j.ecolecon.2012.02.018

Wilson, S. (2022). *Water conservation in Barbados: How GCF is enhancing water security*. Green Climate Fund. https://www.greenclimate.fund/story/water-conservation-barbados-how-gcf-enhancing-water-security. Accessed 9 May 2022.

World Bank. (2006). *Managing water resources to maximize sustainable growth: A country water resources assistance strategy for Ethiopia*. World Bank.

Yeo, S. (2019). Climate finance: The money trail. *Nature, 573*, 328–331.

CHAPTER 5

A Transformative Social Sustainability (Pillar Four)

Abstract Pillar four recognizes that the social dimension is instrumentally important for successful low carbon, sustainability transitions. The social dimension is central to the development of the institutional, community and governance models needed to address our most intractable challenges. Broad and deep social networks are necessary for the successful societal support for deep infrastructure changes. Empowering communities and involving them in decision-making through genuine participatory models can result in policy and planning decisions that are more likely to be acceptable as well as more effective in the long-term. Justice concepts, as guiding principles of policy development, can allow diverse perspectives in decision-making processes, enabling wider reasoning on what is considered legitimate as well as providing a means to inform ethical actions.

Keywords Social dimension · Empowering · Communities · Participatory models · Justice · Ethical

> *Sustainable development is a general idea that must be adapted to specific local social, cultural and political conditions and must be understood and accepted by a large part of society.* (Leal Filho et al., 2016, p. 760)

J. Morrissey and C. P. Heidkamp, *Demanding Sustainability*,
https://doi.org/10.1007/978-3-031-18958-6_5

Recognizing the Social Domain

Though sustainable development as originally framed by the World Commission on Environment and Development (1987) included a clear social mandate, the social dimension has been largely overlooked in the decades since. Abbreviated 'sustainability' discourse has tended to focus on bio-physical environmental issues or has been subsumed into broader discussions dominated by concerns of 'development' and economic growth (Vallance et al., 2011). With the emergence of transitions ideas and a heightened recognition that society-wide behavior change is an important component of any low-carbon transition (see Axon et al., [2018]), the social domain of sustainable development and of transitions, in particular, has garnered increasing attention in recent years.

Studies such as those by Hoppe et al. (2015) and Seyfang and Haxeltine (2012) underline the importance of social sustainability, particularly with regard to local and community-based initiatives. Naber et al. (2017) presents compelling evidence that the social dimension is in fact instrumentally important for transitions, whereby the building of broad and deep social networks is necessary for the successful upscaling of key transition infrastructure such as smart electricity grids. The social dimension is critical in the development of the institutional, community and governance models needed to address humanity's most intractable challenges.

One element of this relates to governance. Success of the SDGs is largely dependent on how effective or not governance arrangements might be (Glass & Newig, 2019). Many deficits are evident in governance practices in relation to sustainable development at present, however, including information deficits and limited knowledge amongst key stakeholders, deficiencies in stakeholder cooperation and engagement and fragmented or uncoordinated policy implementation (Leal Filho et al., 2016). While Irvin and Stansbury (2004) argue that citizen involvement in public decision-making may in fact be ineffective and wasteful in some circumstances, Glass and Newig (2019) identify participation as centrally important, particularly when considering SDG achievement.[1]

[1] Combined with factors such as democratic institutions, economic performance and education (Glass & Newig, 2019).

'Both participatory and democratic governance structures seem to facilitate....acceptance of policies directed towards the achievement of sustainable development' (Glass & Newig, 2019, p. 11). Multiple potential benefits are identified in the literature from empowering communities and engaging them through meaningful participatory models (Morrissey, 2021):

- Participation of diverse societal actors in deliberative and participatory processes can potentially make explicit the difficult choices and planning trade-offs related to long-term complex challenges such as the climate crisis (Cavaleri Gerhardinger et al., 2020).
- Participatory models can be an effective means of acknowledging local needs, values and interests for recognition justice (Williams & Doyon, 2019).
- Enhanced legitimacy and acceptability; decisions arrived at through deliberative and participatory means may be better supported and are likely to be more acceptable to communities.
- By contrast, weak and unequal actor relations can ultimately lead to ineffective environmental governance outcomes, as reported by Thompson (2018).
- Citizens can identify and solve problems, can provide valuable knowledge and they can help address technical as well as social aspects of problems (Sauermann et al., 2020).

Figure 5.1 illustrates the type of governance transformations required. This diagram has been inspired by the figure 'Horizontal, vertical, and hierarchical upscaling in EU governance' presented in Kern (2019).

Panel A illustrates a hierarchal expression of political power, typified by asymmetrical resource and capacity availability, as well as a linear, formalized and distanced relationship between scales, from global to national, city-region, community and finally individual. By contrast, Panel B shows a governance model characterized by aligned networks, typified by principles of alignment, subsidiarity, capacity building and participation. In Panel B, lower tiers on the governance hierarchy are empowered by those tiers which are higher up, with higher levels of governance playing critical mediation, co-ordination and alignment functions, while lower levels of governance are equipped to enact meaningful change at smaller scales.

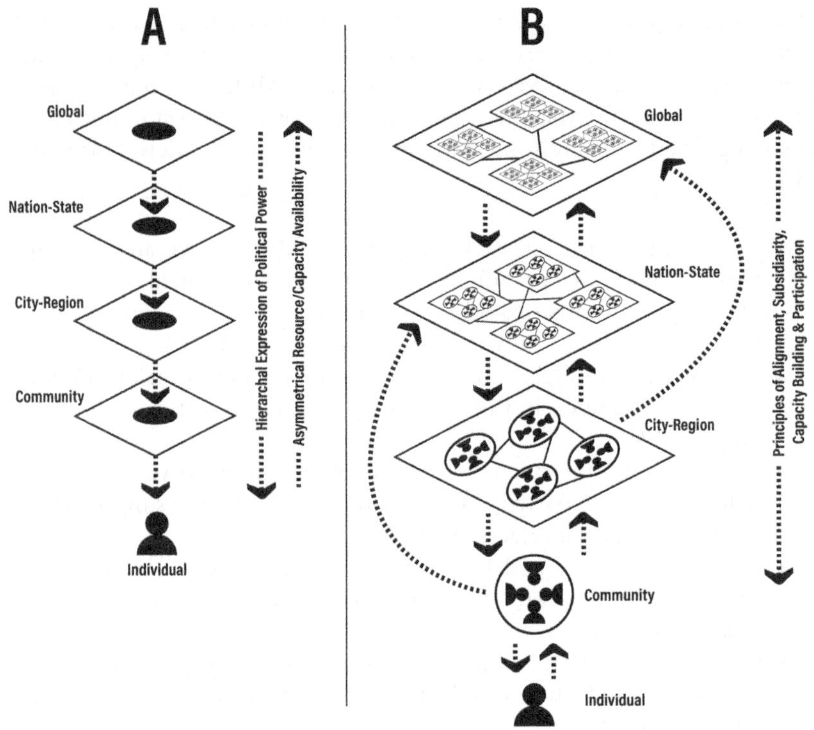

Fig. 5.1 From top-down to integrated, aligned and participatory governance models

> **Box 5.1: Key Data Points—Pillar Four**
> Pillar four calls for a governance model characterized by aligned networks, typified by principles of alignment, subsidiarity, capacity building and participation.

Justice Principles

Justice is fundamental here and key to the successful navigation of the complex decision-making of climate crisis responses, for instance (Morrissey, 2021). Justice concepts can allow diverse perspectives in decision-making processes and can make explicit the social implications of policy alternatives (Bennett, 2018), particularly by enabling wider reasoning on what are considered legitimate and ethical actions (Forsyth, 2014). Jasanoff (2018) emphasizes a move away from traditional top-down command and control hierarchal governance models in a 'humble approach' to just energy transitions. Jasanoff argues specifically for an explicit acknowledgement of procedural justice principles and prioritization of these. The strong public reaction and mobilization evident in the examples of the *Gillet Jeunes* movement in France (2018–2020) and the water protest movement in Ireland (2014–2017) amongst others show the limitations and risks of sustainable development transition policies which are developed in an overtly top-down manner. Popke et al. (2016) warns that climate change vulnerability and inequality challenges may become compounded if resilience approaches lack justice considerations. The case for proactive attention to social justice and inclusive decision-making has significance from both ethical and instrumental perspectives therefore (Bennett, 2018; Grimmel et al., 2019).

Meaningful local responses to the climate crisis, including design of effective climate adaptation policies, require that communities are effectively engaged in discussions about vulnerability, in mapping their own community's vulnerabilities and in co-designing adaptation policies (Holland, 2017). Genuine participatory models can empower communities and can precipitate policy and planning decisions that are more likely to be acceptable and more effective in the long run (Garland et al., 2019). However, embedding participatory models in community resilience building is contingent on genuine capacity building, supported by requisite financial resources (Howath, 2019). Such capacity includes the capacity to engage with (and firstly access) scientific knowledge, the scope to access resources and expertise to enact consensus-derived decisions and the ability to participate in and meaningfully influence local-decision-making (Garland et al., 2019). On this issue, Pasgaard and Dawson (2019) call for governance models which consider local values and identities and importantly which are aligned with a capabilities approach. From the perspective of the research community, future research must define knowledge more broadly and identify whom it empowers. As an academic community, meaningful involvement in delivering solutions to the climate crisis suggests humility in embracing the opportunities presented by the movement towards citizen science. The role of transdisciplinarity is important here too, characterized by collaboration, partnership, stakeholder participation and knowledge co-production. This too calls for humility on the part of academic practitioners and a recognition that no one group, approach, discipline or theory has a 'monopoly on wisdom' or insight.

Box 5.2: Key Data Points—Pillar Four
Justice concepts can enable space for a diversity of perspectives in decision-making processes and can make explicit the social implications of policy alternatives (Bennett, 2018), enabling wider reasoning on the legitimacy and ethicacy of alternative actions (Forsyth, 2014).

Social Innovation

Social Innovation provides one instrument through which novel approaches to social sustainability can be identified, trialed and expanded upon. The OECD (2022) define *social innovation* as 'the design and implementation of new solutions that imply conceptual, process, product, or organizational change, which ultimately aim to improve the welfare and well-being of individuals and communities'. The 'social' dimension of sustainable development is therefore emphasized (Edwards-Schachter & Wallace, 2017), with a concentration on social change and on social practices, as opposed to the primarily profit motivations of technological innovation. Social innovation typically involves novel ways of acting and thinking that change dominant paradigms (Cajaiba-Santana, 2014). There is an emphasis on enhancing bottom-up responses and on integrating social, economic and environmental objectives (Soma et al., 2018). Social innovation, characterized by novel ideas which have diffused, been accepted and have had impact (Alic & Sarewitz, 2016), is therefore simultaneously both a process and an outcome, two aspects which are strongly interwoven (Baker & Mehmood, 2015; Soma et al., 2018).

The OCED (2022) suggests that social innovations can deliver better societal solutions when an enabling policy framework is developed to support the co-construction and implementation of novel ideas and approaches. The European Commission has attempted this with Innovation Union (2010) and Social Investment Package (2013) initiatives, with aims to facilitate uptake and scaling-up of social innovations (European Commission, 2022). The 'Board of Innovation' profile of *50 social impact innovations that might save the world* provides an excellent overview of the diversity, range and focus of both technical and social innovations across community, equality, empowerment and environment categories (Sutton, 2022). Select examples of social innovations are provided in Table 5.1.

Box 5.3: A Transformative Social Sustainability in Action

A TRANSFORMATIVE SOCIAL SUSTAINABILITY IN ACTION: 'Los Pinos: Comunidad del Buen Vivir'

The Los Pinos community, located in the northern Sierra region in Ecuador, provides an inspiring vignette of direct participation in decision-making by a community to develop a neighbourhood based on sustainability principles, or as framed in Ecuador, the 'Buen Vivir' concept. *Buen vivir* is described by Vanhulst and Beling (2014, p. 54) as a 'distinctive Latin American variant' of sustainable development ideas. Drawing on the continent's indigenous cultures, particularly those of Ecuador and Bolivia, *Buen Vivir* suggests a way of living the present in harmony, by assuming and respecting differences and complementarities (amongst humans and between humans and non-humans) in holistic and mutualistic ways (Vanhulst & Beling, 2014). *Buen Vivir* has begun to resonate with a wide range of groups and social movements globally, but especially so in Ecuador and Bolivia.

The Los Pinos community consists of more than 1300 inhabitants, across approximately 300 families of diverse cultural backgrounds. 61.2% of the community are women and 58.9% are between 18 and 44 years old. The community has significant levels of economic deprivation, typified by problems of lack of access to land and basic services and by underemployment. In 2006, the Los Pinos Community Development Committee (CDC-LP) began a process of occupying public land in the Parish of Cutuglagua, which is part of the Mejía Canton, in the Province of Pichincha. The project coalesced around a political struggle characterized by collective actions around the right to housing and deliberate participation in the social production of community (Rodríguez et al., 2016).

The committee (CDC-LP) instigated cooperation agreements with diverse stakeholders, including Architects Without Borders UK (ASF-UK), United Youth Organization (OJU) and Salesian Polytechnic University (UPS), aiming to develop the project of the 'Barrio del Buen Vivir' and to consolidate a network of expertise, mobilization and advocacy around the project (Rodríguez et al., 2016).

A sustained process of community and organizational strengthening facilitated participatory socio-urban design processes, generating spaces for dialogue between technical-scientific knowledge and popular knowledge. The participatory design of community spaces involved (Rodríguez et al., 2016):

- The strengthening of social cohesion, community organization and the sense of community;
- Preventive and promotional approach to environmental and community health;
- Development of productive and cooperative actions based on the popular and solidary economy.

After a period of conflict with bodies such as the National Secretariat for Policy Management (SNGP), the project has since been recognized by the municipality of Mejía (Urbanmonde.org, 2017). The project was awarded the Latin American Habitat Social Production Award (PSH) 2017 (Universidad Politécnica Salesiana (UPS), 2017).

Delivering Pillar Four

Pillar four represents the key pillar through which societal engagement can be mobilized and sufficient support for the four other pillars generated. As described, pillar four requires a step-change from governance models of the past, including command and control, top-down models and the market-driven approaches of the neo-liberal era. Instead, a governance model characterized by subsidiarity, capacity building and participation is necessary. The rationale is that leadership, agency and action is required at all levels of society to adequately respond to our linked economic, social and ecological crises. This is best achieved by fostering and empowering engaged citizens as opposed to passive consumers. The following reforms may need to be demanded from incumbent regimes, rather than be willingly granted:

Table 5.1 Selected social impact innovations[2]

Social innovation	*Location*	*Url*	*About*
CitizenLab	Brussels, EU	www.citizenlab.co	CitizensLab is a digital citizen participation platform for governments which provides a digital participation platform to communicate with citizens on local topics and include them in decision-making. CitizenLab's mission is to 'build stronger democracies by making public decision-making more inclusive, participatory and responsive'
IFIXIT	Online community	eustore.ifixit.com	IFIXIT is pushing for the 'right to fix' products with the aim to 'ditch the throwaway economy', by sharing online repair manuals and resources to help people fix or modify everyday items such as phones, vehicles and home appliances. As of May 2022, 82,018 free online manuals were listed on the IFIXIT website
Triodos	UK	www.triodos.com	Triodos is a financial institution that lends money to organizations only when they can prove a positive impact on society and the environment. They also aim to be more transparent and open about their charges to make banking more honest. Triodos reports a headline figure of 851 kilo-tonnes CO_2e in avoided emissions as a result of renewable energy loans and investments in 2021

(continued)

[2] After Sutton (2022).

Table 5.1 (continued)

Social innovation	*Location*	*Url*	*About*
Teacherly	UK/online community	teacherly.io	Teacherly aims to create a global online community of teachers dedicated to improving learning outcomes. They achieve this by sharing content, best practices and professional development materials. The platform allows teachers to collaborate with other teachers on lessons, to prepare lessons with interactive tools and to track single student progress through the use of dedicated analytical tools

- Policy acknowledgement of the importance of broad-based social networks for low-carbon transition, followed by the development of appropriate governance mechanisms to empower and facilitate these networks
- As with other pillars, mobilization of a national dialogue on sustainability, resilience and prosperity. For pillar four, this would entail a national discussion on the best models, frameworks and structures through which national dialogues can fairly, transparently, inclusively and effectively be conducted
- Meaningful application of subsidiarity with regard to energy, land use and food system planning
- Concerted local and regional scale capacity building, including appropriate provision of financial resources to support this
- Capacity building of citizens at the individual level, including free education (educational loan debt forgiveness) and training campaigns on scientific literacy, citizen science and on local, practical responses to the climate crisis
- Re-calibration of academic practice towards community and practice-based research
- Co-creation of knowledge in a transdisciplinary environment, via partnership and knowledge co-production without limiting stakeholder participation
- More strident civil society campaigns, including more resolute political campaigning, targeted application of obstruction and boycott tactics, civil disobedience and non-participation in the consumer economy **where blatant climate injustice is evident**.

References

Alic, J. A., & Sarewitz, D. (2016). Rethinking innovation for decarbonizing energy systems. *Energy Research & Social Science, 21*, 212–221. https://doi.org/10.1016/j.erss.2016.08.005

Axon, S., Morrissey, J., Aiesha, R., et al. (2018). The human factor: Classification of European community-based behaviour change initiatives. *Journal of Cleaner Production, 182*, 567–586. https://doi.org/10.1016/j.jclepro.2018.01.232

Baker, S., & Mehmood, A. (2015). Social innovation and the governance of sustainable places. *Local Environment, 20*, 321–334. https://doi.org/10.1080/13549839.2013.842964

Bennett, N. J. (2018). Navigating a just and inclusive path towards sustainable oceans. *Marine Policy, 97*, 139–146. https://doi.org/10.1016/j.marpol.2018.06.001

Cajaiba-Santana, G. (2014). Social innovation: Moving the field forward. A conceptual framework. *Technological Forecasting and Social Change, 82*, 42–51. https://doi.org/10.1016/j.techfore.2013.05.008

Cavaleri Gerhardinger, L., Martins de Andrade, M., Ribeiro Correa, M., & Turra, A. (2020). Crafting a sustainability transition experiment for the Brazilian blue economy. *Marine Policy, 120*. https://doi.org/10.1016/j.marpol.2020.104157

Edwards-Schachter, M., & Wallace, M. L. (2017). 'Shaken, but not stirred': Sixty years of defining social innovation. *Technological Forecasting and Social Change, 119*, 64–79. https://doi.org/10.1016/j.techfore.2017.03.012

European Commission. (2022). Social innovation. In: European industrial strategy. https://ec.europa.eu/growth/industry/strategy/innovation/social_en. Accessed 12 May 2022.

Forsyth, T. (2014). Climate justice is not just ice. *Geoforum, 54*, 230–232. https://doi.org/10.1016/j.geoforum.2012.12.008

Garland, M., Axon, S., Graziano, M., Morrissey, J., & Heidkamp, C. P. (2019). The blue economy: Identifying geographic concepts and sensitivities. *Geography Compass*, 1–21. https://doi.org/10.1111/gec3.12445

Glass, L.-M., & Newig, J. (2019). Governance for achieving the Sustainable Development Goals: How important are participation, policy coherence, reflexivity, adaptation and democratic institutions? *Earth System Governance, 2*, 100031. https://doi.org/10.1016/j.esg.2019.100031

Grimmel, H., Calado, H., Fonseca, C., & Suárez De Vivero, J. L. (2019). Integration of the social dimension into marine spatial planning—Theoretical aspects and recommendations. *Ocean and Coastal Management, 173*, 139–147. https://doi.org/10.1016/j.ocecoaman.2019.02.013

Holland, B. (2017). Procedural justice in local climate adaptation: Political capabilities and transformational change. *Environmental Politics, 26*, 391–412. https://doi.org/10.1080/09644016.2017.1287625

Hoppe, T., Graf, A., Warbroek, B., et al. (2015). Local governments supporting local energy initiatives: Lessons from the best practices of Saerbeck (Germany) and Lochem (The Netherlands). *Sustain, 7*, 1900–1931. https://doi.org/10.3390/su7021900

Howath, C. (2019). *Resilience to climate change*. Palgrave Macmillon.

Irvin, R. A., & Stansbury, J. (2004). Citizen participation in decision making: Is it worth the effort? *Public Administration Review, 64*, 55–65.

Jasanoff, S. (2018). Just transitions: A humble approach to global energy futures. *Energy Research & Social Science, 35*, 11–14. https://doi.org/10.1016/j.erss.2017.11.025

Kern, K. (2019). Cities as leaders in EU multilevel climate governance: Embedded upscaling of local experiments in Europe. *Environmental Politics, 28*, 125–145. https://doi.org/10.1080/09644016.2019.1521979

Leal Filho, W., Platje, J., Gerstlberger, W., et al. (2016). The role of governance in realising the transition towards sustainable societies. *Journal of Cleaner Production, 113*, 755–766. https://doi.org/10.1016/j.jclepro.2015.11.060

Morrissey, J. E. (2021). Coastal communities, blue economy and the climate crisis: Framing just disruptions. *Geographical Journal*. https://doi.org/10.1111/geoj.12419

Morrissey, J., & Heidkamp, C. P. (2022). Sustainability after Covid: Pillars for a just transition. *Environmental Sustainability*. https://doi.org/10.1007/s42398-022-00231-y

Naber, R., Raven, R., Kouw, M., & Dassen, T. (2017). Scaling up sustainable energy innovations. *Energy Policy, 110*, 342–354. https://doi.org/10.1016/j.enpol.2017.07.056

OECD. (2022). Social innovation. In: Local economic and employment development. LEED program. https://www.oecd.org/regional/leed/social-innovation.htm#:~:text=Social-innovation-refers-to-the-wellbeing-of-individuals-and-communities. Accessed 11 May 2022.

Pasgaard, M., & Dawson, N. (2019). Looking beyond justice as universal basic needs is essential to progress towards 'safe and just operating spaces.' *Earth System Governance, 2*, 100030. https://doi.org/10.1016/j.esg.2019.100030

Popke, J., Curtis, S., & Gamble, D. W. (2016). A social justice framing of climate change discourse and policy: Adaptation, resilience and vulnerability in a Jamaican agricultural landscape. *Geoforum, 73*, 70–80. https://doi.org/10.1016/j.geoforum.2014.11.003

Rodríguez, M., Grondona-Opazo, G., Erazo, J., & Festjens, J. (2016). Disputas urbano-populares: creatividad y antagonismos para la construcción de barrios del Buen Vivir en Quito, Ecuador. In: P. Abramo, M. R. Mancilla, & J. E. Espinosa (Eds.), *Ciudades populares en disputa ¿Acceso a suelo urbano para todos?* (1st edn.). Ediciones Abya-Yala - CLACSO - Universidad Federal de Río de Janeiro - Universidad Politécnica Salesiana.

Sauermann, H., Vohland, K., Antoniou, V., et al. (2020). Citizen science and sustainability transitions. *Research Policy, 49*, 103978. https://doi.org/10.1016/j.respol.2020.103978

Seyfang, G., & Haxeltine, A. (2012). Growing grassroots innovations: Exploring the role of community-based initiatives in governing sustainable energy transitions. *Environment and Planning C: Government and Policy, 30*, 381–400. https://doi.org/10.1068/c10222

Soma, K., Van Den, B. S. W. K., Hoefnagel, E. W. J., et al. (2018). Social innovation—A future pathway for Blue growth? *Marine Policy, 87*, 363–370. https://doi.org/10.1016/j.marpol.2017.10.008

Sutton, A. (2022). 50 social impact innovations that might save the world. In: Board innovation. https://www.boardofinnovation.com/blog/50-social-impact-innovations-helping-to-save-the-world/. Accessed 11 May 2022.

Thompson, B. S. (2018). The political ecology of mangrove forest restoration in Thailand: Institutional arrangements and power dynamics. *Land Use Policy, 78*, 503–514.

Universidad Politécnica Salesiana (UPS). (2017). Social habitat 2017 rewarded the project "Los Pinos: Comunidad del Buen Vivir." https://www.ups.edu.ec/en/noticias?articleId=9465957. Accessed 3 Mar 2022.

Urbanmonde.org. (2017). Community Los Pinos (Ecuador). In: Social production of habitat award 2017. https://www.youtube.com/watch?v=OMdIkY9Z8x8. Accessed 1 Mar 2022.

Vallance, S., Perkins, H. C., & Dixon, J. E. (2011). What is social sustainability? A clarification of concepts. *Geoforum, 42*, 342–348. https://doi.org/10.1016/j.geoforum.2011.01.002

Vanhulst, J., & Beling, A. E. (2014). *Buen vivir: Emergent discourse within or beyond sustainable development?*. Elsevier B.V.

Williams, S., & Doyon, A. (2019). Justice in energy transitions. *Environmental Innovation and Societal Transitions, 31*, 144–153. https://doi.org/10.1016/j.eist.2018.12.001

World Commission on Environment and Development (WCED). (1987). *Our common future*. Oxford University Press.

CHAPTER 6

A Just Resilience (Pillar Five)

Abstract Pillar five a just resilience is framed in honest recognition of the hard times to come. If pillar one established the boundaries, pillar five seeks to protect those boundaries. Even if a warming limit of 1.5 °C were to be achieved, the impacts and consequences for communities globally will be severe in the decades to come, with many communities already grappling with climate-related devastation. The just disruptions ideas presented under pillar five center the application of justice principles and social sustainability in community responses to disruptions of socio-economic and environmental systems.

Keywords Justice · Resilience · Social sustainability · Community responses · Just disruptions

> *The scale of the threats to the biosphere and all its lifeforms—including humanity—is in fact so great that it is difficult to grasp for even well-informed experts.* (Bradshaw et al., 2021, p. 1)

J. Morrissey and C. P. Heidkamp, *Demanding Sustainability*,
https://doi.org/10.1007/978-3-031-18958-6_6

Existential Risks[1]

The carbon budget remaining to ensure planetary heating above pre-industrial levels is limited to 1.5 °C is ~400 $GtCO_2$ (UNEP, 2021).[2] In May 2022, the World Meteorological Organization (WMO) warned of a 50:50 chance that the annual average global temperature would temporarily reach 1.5 °C above pre-industrial levels for at least one of the next five years (to 2027), with this probability increasing over time (World Meteorological Organization [WMO], 2022). The likely scenario of global heating of 2 °C magnitude is likely to increase exposure to climate-related risks and poverty for several hundred million people by 2050 (Xu et al., 2019). The risks are existentially challenging, with interacting socio-economic and ecological dimensions; stresses to prosperity, human health and well-being will likely erode political capacity and the ability to protect vital ecosystem services. Gleeson (2018, p. 40) warns 'the tsunami of human and global change will overtake well intentioned action. The scale of the problem will continue to magnify both in terms of threat and species exposure'.

Times of crisis (such as the Covid-19 pandemic), provide narrow windows of opportunity when system transformation potentials become more viable (Blowers, 1997). However, sustainable development policy responses in the decades since the publication of the World Commission on Environment and Development (Brundtland Commission) (1987) Our Common Future report has been largely based on a 'weak sustainability' paradigm, has largely assumed the political-economic structure and has aimed for a slow incrementalism and (socio)-technical transition rather than a more fundamental system transformation; the foundational assumption has been that change is to be achieved gradually and institutional adaptation is to be achieved through consensus (Blowers, 1997). The scale and pace of this incrementalism have been inadequate in the face of accumulating socio-ecological challenges. Notably, the UNEP reports that current commitments expressed in the Nationally Determined Contributions (NDCs) committed to in the Paris Agreement are

[1] Part of this chapter has been edited from work originally published in: Morrissey, J. (2021) Coastal communities, blue economy and the climate crisis: Framing just disruptions, The Geographical Journal, © John Wiley and Sons and is used by permission.

[2] Based on IPCC estimates and 66% chance.

inadequate to reduce emissions sufficiently to deliver a 'safe operating space'. A warming level of over 1.5 °C will be unavoidable if NDC ambitions are not increased radically before 2030 (UNEP, 2019).

Box 6.1: Key Data Points—Pillar Five
There is a 50:50 chance that annual average global temperature temporarily reaches 1.5 °C above pre-industrial levels for at least one of the next five years (to 2027) (World Meteorological Organization [WMO], 2022).

As discussed by Klein (2019), extending the *status quo* unchanged into the future is not a viable option for societies. Bradshaw et al.'s (2021, p. 6) dire warnings of 'a ghastly future' defined by resource conflicts, mass extinction, declining health and climate disruption upheavals are a reminder of the stakes. Younger generations are already reacting with anger and disbelief:

> 'How dare you! You have stolen my dreams and my childhood with your empty words. And yet I'm one of the lucky ones. People are suffering. People are dying. Entire ecosystems are collapsing. We are in the beginning of a mass extinction, and all you can talk about is money and fairy tales of eternal economic growth. How dare you!'
>
> Great Thunberg, Speech Delivered at the U.N. Climate Action Summit, 2019, reported in NPR (2019).

The rapidly evolving climate crisis will by its nature necessitate radical overhaul and change in institutional and governance structures. Folke (2006), pointing to the multi-scalar complexities of social-ecological systems, argues that change is a process to be navigated rather than an obstacle to be avoided. However, given that considerable disruption is likely to be externally imposed from a rapidly changing, unpredictable and increasingly hostile environment, institutional integrity and resilience forms a key tenet of a successful transition. We are therefore faced with the seemingly contradictory challenge of needing to radically reinvent our institutional structures to face the coming challenges, while simultaneously imbuing these structures with requisite resilience to support the fundamentals of societal functioning through the disruptions to come. Halpern and Mason's (2015) 'radical incrementalism' based on multiple small improvements accumulating into bigger impacts may help here. However, only by realistically facing up to the colossal challenges facing humanity will we be able to chart a course to a less-ravaged future (Bradshaw et al., 2021).

Box 6.2: Key Data Points—Pillar Five
We are faced with the seemingly contradictory challenge of needing to radically reinvent our institutional structures to face the coming challenges, while simultaneously imbuing these structures with requisite resilienceto support the fundamentals of societal functioning through the disruptions to come.

'Bounce Forward' Sustainability

Resilience thinking has shifted from original 'return to the original state' (bounce back) understandings towards recognition that adaptation and transformation can be used to further certain social goals (bounce forward) (Copeland et al., 2020; Weichselgartner & Kelman, 2015). Social justice considerations are key. Here we present a 'just disruptions' framework, which seeks to integrate just transitions and resilience perspectives, originally published in Morrissey (2021). A 'just disruptions' approach to development seeks to recognize that ecological, socio-economic and political domains are in flux and that the socio-economic context cannot be assumed. Just Disruptions are defined as *the means through which communities respond to disruptions of socio-economic and environmental systems, so as to enhance social sustainability through the appropriate application of justice principles* (Morrissey, 2021, p. 7).

The impact of climate-related disruptions on sustainability transitions processes, emerging low-carbon economies and overall socio-ecological sustainability is represented in Fig. 6.1. Three scenarios of post-disruption sustainability transition dynamics are illustrated, characterized as 'bound forward', 'bounce back' and 'bounce backwards'.

A 'bounce forward' (Scenario A) recovery posits that sustainability transition dynamics are accelerated post-disruption and that the economy becomes more sustainable, across ecological, social and economic domains (see Manyena et al., 2011). Under this scenario, disruption serves as an impetus to rapidly develop a sustainable, resilient and just

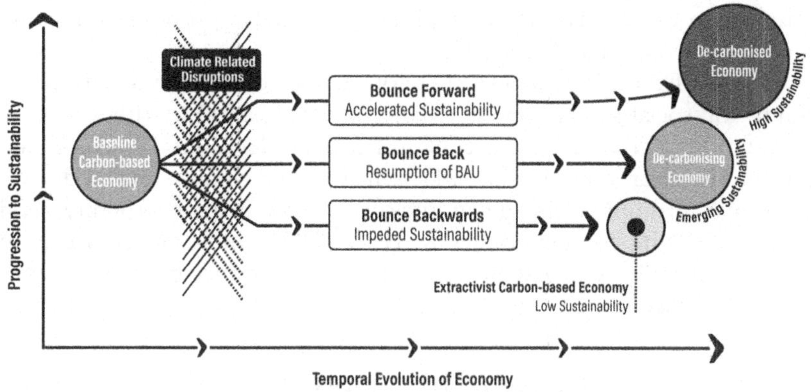

Fig. 6.1 Impact of climate-related disruptions on transitions processes[3]

economy. Scenario A frames disruption, therefore, as an opportunity for enhanced sustainability, including social enhancement, rather than as a simple return to the *status quo ante* (Manyena et al., 2011).

Bounce forward represents an evolution from original understandings of resilience as a 'return to the original state', that is a 'bounce back' (Scenario B) recovery. Under Scenario B, sustainability transition dynamics resume to proceed as previously (Weichselgartner & Kelman, 2015). Scenario B assumes that the Business-as-Usual (BAU) trajectory for economies is one towards slowly developing low-carbon economies, with varying levels of emerging sustainability. Under this scenario, post-disruption recovery is quickly achieved and the *status-quo ante* is restored.

Finally, a 'bounce backwards' (Scenario C) recovery sees sustainability transition dynamics slowed and even reversed, with climate-related disruptions initiating a reversal in overall sustainability and negatively influencing the momentum of sustainability transitions. Recent transitions research has started to explore these issues, including the study by Boons et al. (2021) on the impact of the disruption of the Covid-19 pandemic on transition processes in the UK and work by Chlebna and Mattes (2020) on the fragility of regional energy transitions. Under Scenario C, progress towards BE is significantly impeded in the aftermath of climate-related disruption, with a post-disruption context of crisis precipitating

[3] First published in Morrissey (2021).

a return to extractivist and ecologically exploitative economies, moving further from sustainability ideals.

'Bounce forward' resilience therefore frames disruption as an opportunity for social enhancement (Manyena et al., 2011), with a central recognition of the importance of justice considerations (Popke et al., 2016). However, considerable risks remain; while climate-related disruptions may prompt a change in socio-economic system structures and norms (Institute for Economics & Peace, 2020), it is far from certain whether resulting change dynamics will result in 'bounce back' or 'bounce forward' recoveries.

Targeting 'Just Disruptions'

Figure 6.2 outlines how just disruptions thinking can be applied to inform transitions from business as usual economy models to models of regionally embedded, resilient, just and sustainable decarbonized economies. Economies characterized by business as usual economics are shown to

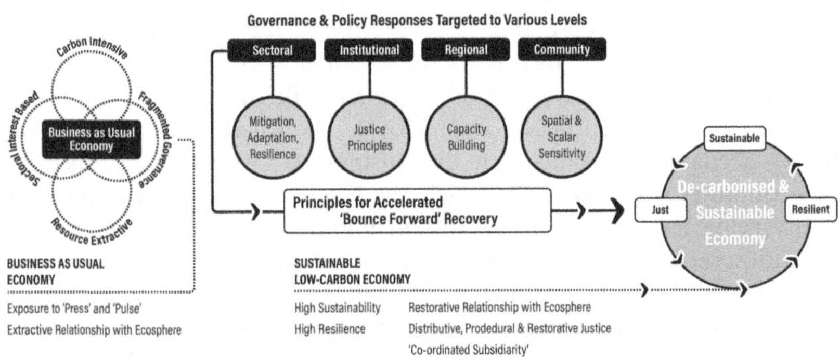

Fig. 6.2 A Just disruptions framework for sustainable, just and resilient low carbon economies[5]

the left side of the diagram. These economies are characterized as extractive, resource-intensive, carbon dependent and typified by fragmented governance arrangements.[4]

By contrast, the decarbonized and sustainable economy is typified by high levels of resilience, a restorative relationship with the ecosphere and by deliberative and participatory models of governance, typified by distributive, procedural and restorative justice. The four facilitating principles of the framework, shown in the middle of the diagram in Fig. 6.2, represent the essential strategies through which the transition from the business as usual economy models to decarbonized and sustainable economy models can be realized. These principles are framed in recognition of the imperfect conditions under which transitions are likely to occur, i.e., in the context of ongoing climate disruptions (of both 'press' and 'pulse' nature), a paucity of economic resources for communities and the progressive degradation of global and regional ecosystems.

The facilitating principles are integration of policy and governance responses to the climate crisis across mitigation, adaptation and resilience building; rights-based and capabilities-based approaches to justice, including principles of distributive, procedural and restorative justice; capacity building and recognition of the realities of socio-spatial and scalar differentiation in the development of sustainable, low-carbon economies. The outlined principles provide the means through which the path dependency and inertia of current BAU economies could be transformed into more sustainable, just and resilient economies. In particular, capacity building and the application of justice principles may be understood as facilitating and empowering, process-related concepts rather than end-point 'destinations'. The appropriate application of these principles may have instrumental value to propogate sustainability and initiate viable pathways to the emergence of sustainable, just and resilient economies. The application of justice principles can help with the challenges of complex decision-making in this context (Morrissey, 2021). Climate change policy measures requires broad public support to be politically feasible and socially acceptable (Moser & Pike, 2015). Justice principles

[4] For example, the management of coastal zones has traditionally been conducted in a fragmented manner, with responsibilities for coastal activities being spread across many agencies, according to Berkes (2015).

[5] First published in Morrissey (2021).

can precipitate diverse perspectives in decision-making processes and can make the social implications of policies explicit and formally recognized (Bennett, 2018).

Box 6.3: Key Data Points—Pillar Five
Capacity building and the application of justice principles must be understood as facilitating and empowering, process-related instruments rather than end-point 'destinations'.

Box 6.4: A Just Resilience in Action
A JUST RESILIENCE IN ACTION:
Scotland's 'green recovery and just transition to net zero'
Scotland updated its *Climate Change Plan 2018–2032*. In December 2020, to introduce more than 100 new policies and proposals to better support green recovery and strengthen transitions dynamic in the wake of the Covid-19 pandemic (The Climate Group, 2021). According to Roseanna Cunningham, Cabinet Secretary for Environment, Climate Change and Land Reform (16 December 2020) *'Our commitment to tackling the twin-crises of climate change and biodiversity loss is unwavering and is central to our green recovery from the COVID-19 pandemic'*. (Scottish Government, 2020). The updated plan now outlines pathways to meeting Scotland's emissions reduction targets to 2032. As well as this, the updated plan explicitly mentions just transitions and aims to provide strategic direction for a so-called green recovery from the disruptions caused by Covid-19 (The Climate Group, 2021).

Highlights include (Scottish Government, 2020):

- £180 million towards an Emerging Energy Technologies Fund (EETF), targeted to support Scottish hydrogen and Carbon Capture and Storage (CCS) industries
- £120 million for zero-emission buses
- £50 million to provide sustainable transport links, or 'Active Freeways'
- £50 million to transform vacant and derelict land
- Reducing overall car-kilometers travelled by 20% by 2030
- 1 million zero-emission homes by 2030
- Enhance biodiversity and create green jobs through peatland restoration and woodland creation.

To support delivery of the plan, an extended Public Engagement Strategy was developed to 'engage, support and encourage climate action' (Scottish Government, 2020). A notable feature is also the development of a Climate Emergency Skills Action Plan, aimed to increase so-called green jobs and access to these jobs through targeted training and reskilling programme (The Climate Group, 2021). While there are criticisms of 'just transitions' in the Scotland context (see Heffron and Mccauley's (2022) discussion of the just transitions framing of support for oil and gas industries to the tune of £62 million announced in June 2020), the central framing of justice in the Scottish Climate response together with the willingness and ability to enact rapid policy change in the face of major societal disruption, while maintaining a central prioritization on climate action, are noteworthy and significant.

> *We will use our recovery from this crisis to build a greener, fairer, and more equal society and economy, with a laser focus on creating new, high-quality, green jobs.* (Forbes, 2020, p. 10)

Delivering Pillar Five

This volume was finalized in the summer of 2022, a year of temperature records and heat waves in Europe, heat waves in India and Pakistan and the US as well as drought in the Horn of Africa and extreme floods in South Africa. It is becoming obvious that, regardless of how successful or not mitigation efforts will be in the time to 2030, the impacts and consequences of a changed climate will be severe in the decades to come, and devastating for many. Pillar five is framed in recognition of this reality and in acknowledgement that, at the local level, adaptation planning needs to begin in earnest worldwide. Pillar five posits justice principles and social sustainability as critical in this planning, requiring a number of dedicated policy responses as follows:

- Policy acknowledgement of the importance of climate adaptation and resilience building, followed by development of appropriate governance mechanisms to empower aligned adaptation responses at difference scales
- As with other pillars, mobilization of a national dialogue on climate adaptation, resilience and vulnerability

- National, regional and local vulnerability assessments, across ecological, economic and social domains
- As with pillar four, concerted local and regional scale capacity building, including appropriate provision of financial resources to support this
- As with pillar one, national, regional and local dialogues on the direction of societal development for prosperity and resilience, framed by the empirical evidence of both the planetary boundaries and the SDGs
- Establishment of governance mechanisms and criteria for strategic and managed retreat; managed retreat involves a decisive move away from those areas or sectors where vulnerabilities and costs are deemed to be excessively burdensome on society, a process which needs to be strongly framed by justice principles
- As with pillar one, a socio-cultural shift from a focus on unconstrained consumerism to a focus on sufficiency, resilience and ecological restoration
- A shift from economic models focused on 'just-in-time' and the primacy of economic efficiency to one more aligned with prudent risk management and 'just-in-case' thinking.

References

Bennett, N. J. (2018). Navigating a just and inclusive path towards sustainable oceans. *Marine Policy, 97*, 139–146. https://doi.org/10.1016/j.marpol.2018.06.001

Berkes, F. (2015). *Coasts for people: Interdisciplinary approaches to coastal and marine resource management*. Routledge.

Blowers, A. (1997). Environmental policy: Ecological modernisation or the risk society? *Urban Stud, 34*, 845–871.

Boons, F., Doherty, B., Köhler, J., et al. (2021). Disrupting transitions: Qualitatively modelling the impact of Covid-19 on UK food and mobility provision. *Environmental Innovation and Societal Transitions, 40*, 1–19. https://doi.org/10.1016/j.eist.2021.04.003

Bradshaw, C. J. A., Ehrlich, P. R., Beattie, A., et al. (2021). Underestimating the challenges of avoiding a ghastly future. *Frontiers in Conservation Science, 1*, 1–10. https://doi.org/10.3389/fcosc.2020.615419

Chlebna, C., & Mattes, J. (2020). The fragility of regional energy transitions. *Environmental Innovation and Societal Transitions, 37*, 66–78. https://doi.org/10.1016/j.eist.2020.07.009

Copeland, S., Comes, T., Bach, S., et al. (2020). Measuring social resilience: Trade-offs, challenges and opportunities for indicator models in transforming societies. *International Journal of Disaster Risk Reduction, 51*, 101799. https://doi.org/10.1016/j.ijdrr.2020.101799

Folke, C. (2006). Resilience: The emergence of a perspective for social—Ecological systems analyses. *Global Environmental Change, 16*, 253–267. https://doi.org/10.1016/j.gloenvcha.2006.04.002

Forbes, K. (2020). Building a net zero economy that is fair for all. In H. Murphy (Ed.), *Pathway to COP26—The role of green finance* (pp. 10–13). The Social Market Foundation.

Gleeson, B. J. (2018). A dangerous transition to hope. In T. Moore, F. de Haan, R. Horne, & B. J. Gleeson (Eds.), *Urban sustainability transitions* (pp. 35–49). Springer.

Halpern, D., & Mason, D. (2015). Radical incrementalism. *Evaluation, 21*, 143–149. https://doi.org/10.1177/1356389015578895

Heffron, R. J., & Mccauley, D. (2022). The 'just transition' threat to our energy and climate 2030 targets. *Energy Policy, 165*, 112949. https://doi.org/10.1016/j.enpol.2022.112949

Institute for Economics & Peace. (2020). *Ecological threat register 2020: Understanding ecological threats, resilience and peace*. Institute for Economics & Peace.

Klein, N. (2019). *On fire: The (burning) case for a green new deal*. Penguin Books Ltd.

Manyena, B., O'Brien, G., O'Keefe, P., & Rose, J. (2011). Disaster resilience: A bounce back or bounce forward ability? *Local Environment The International Journal of Justice and Sustainability, 16*, 417–424. https://doi.org/10.1080/13549839.2011.583049

Morrissey, J. E. (2021). Coastal communities, blue economy and the climate crisis: Framing just disruptions. *Geographical Journal*. https://doi.org/10.1111/geoj.12419

Moser, S. C., & Pike, C. (2015). Community engagement on adaptation: Meeting a growing capacity need. *Urban Climate, 14*, 111–115. https://doi.org/10.1016/j.uclim.2015.06.006

NPR. (2019). Transcript: Greta Thunberg's speech at the U.N. climate action summit. In: Environment. https://www.npr.org/2019/09/23/763452863/transcript-greta-thunbergs-speech-at-the-u-n-climate-action-summit?t=1655156951841. Accessed 1 Apr 2022.

Popke, J., Curtis, S., & Gamble, D. W. (2016). A social justice framing of climate change discourse and policy: Adaptation, resilience and vulnerability

in a Jamaican agricultural landscape. *Geoforum, 73*, 70–80. https://doi.org/10.1016/j.geoforum.2014.11.003
Scottish Government. (2020). Steering Scotland's pathway to net zero. Steering Scotland's pathway to net zero - gov.scot. www.gov.scot. Accessed 2 Apr 2022
The Climate Group. (2021). *Building back greener—Scotland's Climate Change Plan: A path to green recovery*. The Climate Group.
UNEP. (2019). *Emissions Gap Report 2019*. United Nations Environment Programme.
UNEP. (2021). *Emissions Gap Report 2021: The heat is on—A world of climate promises not yet delivered—Executive summary*. UNEP.
Weichselgartner, J., & Kelman, I. (2015). Geographies of resilience: Challenges and opportunities of a descriptive concept. *Progress in Human Geography, 39*, 249–267. https://doi.org/10.1177/0309132513518834
World Commission on Environment and Development (WCED). (1987). *Our common future*. Oxford University Press.
World Meteorological Organization (WMO). (2022). WMO update: 50:50 chance of global temperature temporarily reaching 1.5°C threshold in next five years. In: Press Release Number 09052022. https://public.wmo.int/en/media/press-release/wmo-update-5050-chance-of-global-temperature-temporarily-reaching-15%C2%B0c-threshold. Accessed 3 Mar 2022
Xu, C., Kohler, T. A., Lenton, T. M., et al. (2019). Future of the human climate niche. *Proceedings of the National Academy of Sciences, 117*, 11350–11355. https://doi.org/10.1073/pnas.1910114117

CHAPTER 7

Implementing the Pillars: Towards Shared Sustainable Prosperity

Abstract This chapter discusses a number of themes that cut across all 5 pillars, including: Integrated responses to the climate crisis are required so that transitions initiatives acknowledge both the global imperative for emissions reduction /mitigation together with regional and local imperatives for adaptation; Justice should be considered as a core instrument and driver, rather than aspirational end-point, of sustainability transition. Justice need not be simply an ideal of sustainable economies, emerging at some unknown point into the future, but rather a means or process through which sustainable economies can be realized in actuality; a capabilities approach to climate justice can better frame impacts of adaptation and mitigation actions, while acknowledging socio-spatial differentiation and vulnerabilities; fundamental geographical concepts such as scale and connection, governance and power relations, proximity and distance and space and place are fundamental to understanding processes of transition, vulnerability and resilience building.

Keywords Transitions · Justice · Capabilities · Socio-spatial differentiation · Vulnerabilities · Resilience

J. Morrissey and C. P. Heidkamp, *Demanding Sustainability*,
https://doi.org/10.1007/978-3-031-18958-6_7

I like to envision the whole world as a jigsaw puzzle ... If you look at the whole picture, it is overwhelming and terrifying, but if you work on your little part of the jigsaw and know that people all over the world are working on their little bits, that's what will give you hope. Jane Goodall

Linking the Pillars

Chapters 2–6 have presented 5 'pillars' through which to (re-)build a shared prosperity in the aftermath of the Covid-19 global shock. In actual fact, this has to be a new prosperity, developed with a firm grounding in the triple bottom line of balanced economic-ecological-societal well-being and with a clear-eyed understanding of the costs, risks and threats from BAU approaches. The five pillars, ***an ecological prosperity (pillar one), a decarbonized economy (pillar two), a shared (cost) burden (pillar three), a transformative social sustainability (pillar four)*** and ***a just resilience (pillar five)***, aim to provide a framework through which policymakers, decision-makers, politicians, community groups and the corporate sphere might begin to consider, map out and plan for just transitions in their domains. Figure 7.1 illustrates how we envisage the pillars interlinking and support one another.

Pillar one, an ecological prosperity provides the framework and boundary for all other pillars, establishing an explicitly defined space for an ecological and just prosperity into the future—one with empirically established 'hard' boundaries, based on a strong sustainability understanding of ecological stewardship and the human rights framing provided by the SDGs. Pillar two, a decarbonized economy presents us with the immediate urgency of the climate crisis. The well-being and sustainable development trajectories of future generations critically depend on our collective efforts to realize pillar two in the present decade. Without meaningful and effective action on pillar two in the short term, our collective horizons may shift from concerns about delivering sustainable prosperity globally, to more fundamental concerns of survival for ever-increasing segments of the global population. In support of pillar two, pillar three a shared (cost) burden and pillar four a transformative social sustainability both acknowledge the imperative for fairness and social cohesion in our responses. Profound distributive injustices underly global patterns of resource use and related carbon emissions. In addition, those with least responsibility for the climate crisis are projected to be most severely impacted in the decades ahead, while those with the

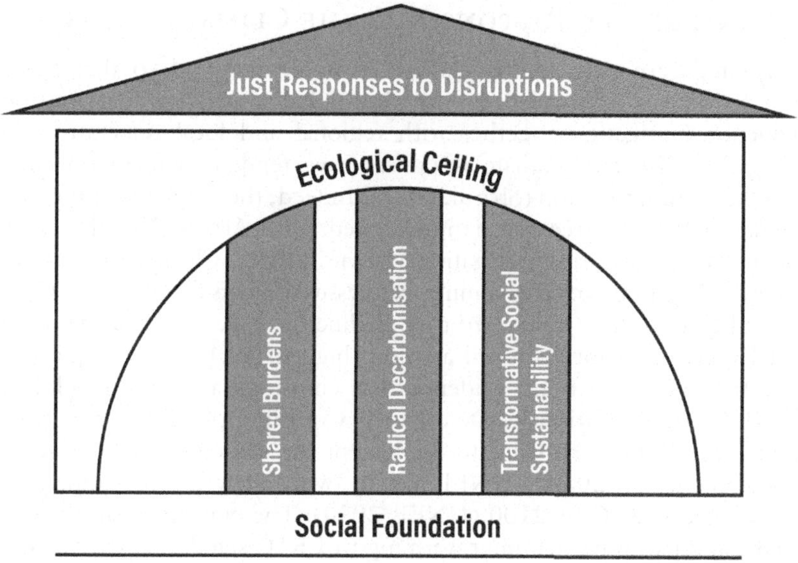

Fig. 7.1 Pillars for shared sustainable prosperity

greatest share of resources continue to produce disproportionate levels of pollution. Those with the resources need to do much more.

Pillar four recognizes that the social dimension is instrumentally important for transitions. Broad and deep social networks are necessary for the successful societal support for, and upscaling of, key transition infrastructure such as smart electricity grids. The social dimension is central to the development of the institutional, community and governance models needed to address our most intractable challenges. Finally, pillar five a just resilience is framed in honest recognition of the hard times to come. If pillar one established the boundaries, pillar five seeks to protect those boundaries. Even if a warming limit of 1.5 °C were to be achieved, the impacts and consequences for communities globally will be severe in the decades to come, with many communities already grappling with climate-related devastation. The just disruptions ideas presented under pillar five center the application of justice principles and social sustainability in community responses to disruptions of socio-economic and environmental systems. As presented in (Chapter 6, Fig. 6.2), a number of themes are cross-cutting across the five pillars, as follows:

Integrated Responses to the Climate Crisis[1]

Integrated responses to the climate crisis are required so that transitions initiatives acknowledge both the global imperative for emissions reduction/mitigation together with regional and local imperatives for adaptation. To date, climate responses have tended to focus on either adaptation or mitigation (Sharifi, 2021). Indeed, the emphasis of the Paris Climate Chance Agreement is imbalanced, with Article 7 of the agreement overtly favouring mitigation (Dovie, 2019). Clearly, the case for urgent mitigation is overwhelming. Successive reports from the IPCC and United Nations have continued to underline the scale of the climate crisis. Total GHG emissions reached a record high of 59.1 GtCO_2-equivalent in 2019 and there is no evidence that emissions are peaking globally (UNEP, 2020). Nationally Determined Contribution (NDCs) commitments under the Paris Climate Agreement are inadequate to bridge the emissions gap in 2030[2] (UNEP, 2020), with likely global temperature rise of at least 3 °C by 2100 (UNEP, 2020). The emissions gap between NDCs and pathways that limit warming to 1.5 °C stands at approximately 28 GtCO_2e per annum (UNEP, 2021).[3]

The scale of this emissions gap and the reality of climate change impacts already evident mean that adaptation is now a necessary inevitability. Despite this, there is a lack of knowledge on how plans or policies can be better integrated for climate change adaptation and community resilience (Kim et al., 2020). Authors such as Kinley (2017) call for a clarification of the parity between mitigation and adaptation,

[1] Part of this chapter has been edited from work originally published in: Morrissey J. (2021) Coastal communities, blue economy and the climate crisis: Framing just disruptions, The Geographical Journal, © John Wiley and Sons and is used by permission.

[2] The emissions gap for 2030 is defined as the difference between global total GHG emissions from least-cost scenarios that keep global warming to 2 °C, 1.8 °C or 1.5 °C (UNEP, 2020).

[3] UNEP point to an emissions reduction potential ~37 GtCO_2e by 2030 (UNEP, 2021), including 12.5 GtCO_2e in Energy, 5.4 GtCO_2e in Industry, 4.7 GtCO_2e in Transport, 1.9 GtCO_2e in Buildings, 5.3 GtCO_2e in Forests, 6.7 GtCO_2e in Agriculture and waste.

enabling, for instance, integration of SDGs to strengthen GHG stabilization and to avoid maladaptation. Similarly, Sharifi (2021) argues that adaptation responses need to be considered at parity with mitigation measures, with potential co-benefits and synergies accruing. Dovie (2019) describes multiple co-benefits of such integration, including development-based natural resource management, biodiversity conservation and maintenance of ecosystem services.

We argue that due regard for the increasing likelihood and severity of climate disruptions is essential, together with an appreciation for the scale of the mitigation challenge and the obligations for urgent action which this challenge puts onto every sector and segment of society. The scale of the crisis is 'of mythic magnitude' as described by Davis (2010) and Gleeson (2018). The public policy response needs to be of equal magnitude, covering mitigation, adaptation and resilience building, across all scales and sectors. However, sustainability and transitions initiatives would likely attract considerably more financial investment within a landscape of wider economic reform and transformation, including the appropriate valuation of ecosystem services such as carbon capture and storage. Those projects providing longer-term carbon sequestration solutions while also offering enhanced community resilience (e.g., coastal protections) should be afforded priority. Furthermore, in the aftermath of disruptive events, community support is likely to be greater for carbon mitigation projects which also build local resilience and adaptive capacity.

Public resources for adaptation, loss and damage need to be urgently scaled up, in addition to the provision of sufficient financial support for mitigation and adaptation (Eckstein et al., 2019). Kim et al. (2020) advocates for proactive planning and public policies to actively build community resilience in the face of such stressors. The level of public debate also needs to be widened and intensified. Herrfahrdt-Pähle et al. (2020) point to the importance of awareness and knowledge building and visionary leadership as key elements of sustainability transformation. The scale of the challenge is unprecedented, 'of mythic magnitude' as described by Davis (2010) and Gleeson (2018). Gleeson (2018, p. 41) continues, '*given that the war we declared upon ourselves has broken out, the arks of human survival will need to be produced like the famous American Liberty ships were in the last great global conflict—rapidly, robustly and with provision for tempest and famine*'.

Box 7.1: Key Data Points—Shared Sustainable Prosperity

> Given that the war we declared upon ourselves has broken out, the arks of human survival will need to be produced like the famous American Liberty ships were in the last great global conflicts—rapidly, robustly and with provision for tempest and famine. Gleeson (2018, p. 41)

Justice as an Instrument of Transition, Sustainability and Resilience

Justice should be considered as a core instrument and driver, rather than aspirational end-point, of sustainability transition. Justice need not be simply an ideal of sustainable economies, emerging at some unknown point into the future, but rather a means or process through which sustainable economies can be realized in actuality. Here the argument is made that justice considerations become ever more important in the context of climate disruptions and associated emergent patterns of vulnerability and disadvantage, as discussed by Schlosberg (2012). While climate-related disruptions may prompt a change in socio-economic system structures and norms (Institute for Economics & Peace, 2020), it is far from certain whether resulting change dynamics will result in 'bounce back' or 'bounce forward' recoveries. In addition, contextual factors, such as poverty levels, development interventions, or management policies are constantly in flux (Owen, 2020), making local resilience building both difficult to define and to track. Further, resilience approaches which are blind to justice considerations risk eliding sources of inequality related to climate change vulnerability in favor of a prioritization of the overall resilience of the system (Popke et al., 2016).

Justice principles can facilitate reasoned public debate on the key question, *'what's the right thing to do?'* (Sandel, 2009). As described by Sandel (2009, p. 261), 'a just society involves reasoning together about the good life'. Justice principles therefore present not simply a means to inform ethical actions but can enable wider reasoning on what is considered legitimate (Forsyth, 2014). Perceptions on fairness and transparency are likely to enhance the legitimacy and community buy-in of sustainability-related projects. Participation of diverse societal actors should ideally make planning trade-offs clearer and better allow for equity considerations in the design of future policies (Cavaleri Gerhardinger et al., 2020).

This is important when considering the difficult choices ahead, particularly with regard to investment options for future sustainability. Which sectors should be prioritized? How is the balance between mitigation and adaptation to be decided? At what scale should responses be targeted at? Particularly important is the consideration of the cost–benefit asymmetry of climate mitigation, whereby costs are highly concentrated spatially and temporally (higher taxes, investment costs, costs of transitions etc. to be borne by present generations) and benefits are diffused across space, globally and over time (avoidance of climate extremes, preservation of favorable environmental conditions etc. to be experienced on inter-generational time-scales).

For this and for other disruption related dilemmas, community buy-in is critical for successful problem-solving. The *'New Blue Economy in Kaikōura'* project in New Zealand provides an excellent example of the potential of participatory processes to first define and subsequently develop regional Blue Economy projects according to sustainability principles. In this project, researchers from the University of Auckland have worked with the Kaikōura community 'to develop a practical, long-term'value proposition' for a sustainable Blue Economy' as a key strategy to foster recovery from the 2016 earthquakes. As a result of this work, Kaikōura has prominent Blue Economy champions and strong stakeholder linkages, as well as a developing platform for Blue Economy activities (Lewis, 2018). There are many benefits to be derived from empowering communities and involving them meaningfully in decision-making through genuine participatory models such as that of *New Blue Economy in Kaikōura*. Policy and planning decisions arrived at in this way are more likely to be acceptable to the community in question and to be more effective in the long-term. Enabling marginalized groups to self-identify their own needs and priorities can also help to mitigate misrepresentation (Fitzgibbons & Mitchell, 2019). An acknowledgement of the diversity of local needs, values and interests is also key for recognition of justice (Williams & Doyon, 2019).

Capacity Building for Resilience

Capacity building, informed by a capabilities approach to justice, is central to meaningful participation. Wood and Roelich (2019) underline the importance of *capabilities*, which can provide a holistic representation of well-being. A capabilities approach to justice focuses on the ability of individuals to convert primary goods into meaningful outcomes in their lives (Brighouse & Robeyns, 2010). Political participation and procedural justice can be represented as capabilities necessary to construct a functioning life (Schlosberg, 2012). This perspective is important as emerging and future impacts of the climate crisis on basic capabilities are significant and are likely to cluster around already disadvantaged individuals and communities (Schlosberg, 2012).

As articulated by Sen (2008, p. 78), capabilities 'would seem to demand attention in any theory of justice and more generally in social assessment, such as the freedom to be well nourished, to live disease free lives, to be able to move around, to be educated, to participate in public life, and so on'. Policy responses cannot be fully effective without due recognition of socio-spatial differentiation of capabilities across society (Catney et al., 2014). Such differentiation of capabilities can inform different socio-cultural contexts. Locally variant values and meanings may be attached to different capabilities in different places and amongst different groups, for instance (Popke et al., 2016; Schlosberg, 2012).

Approaches to resilience building risk embedding or exacerbating social exclusion in the absence of acknowledging diverse experiences of risks or socio-economic barriers to resilience (Forsyth, 2018). Application of a capabilities approach to climate justice can provide a broader perspective and frame through which to understand the impacts of adaptation and mitigation actions (Wilson Furlan Matos Alves and Barberio Mariano, 2018). Further, a capabilities approach inherently fosters recognition of aforementioned socio-spatially specific vulnerabilities (Schlosberg, 2012). As articulated by Holland (2017, p. 398) 'vulnerable citizens must be able to exert some control over either the content of what adaptation protects, or over the decision rules through which such content is determined'.

Likewise, the socio-economic health of many communities depends on empowering people and groups to make change locally, in response to locally specific climate change impacts and emergent dynamics of economic change (Garland et al., 2019; Morrissey & Heidkamp, 2022). This includes the capacity to access and engage with scientific knowledge, the ability to participate in local decision-making and the scope to realize consensus-derived decisions in practice (Garland et al., 2019; Morrissey & Heidkamp, 2022). Communication, collaboration and co-production are elementary in building resilience to climate shocks, Table 7.1 (Howath, 2019).

These elements are consistent with principles of procedural justice. At the community level, processes in which local citizens can obtain sufficient information, have the ability to express their opinions and have their voices heard on an equal footing (Bowles et al., 2019) constitute basic building blocks for procedural justice. Appropriate decision-making arenas are then required for deliberation on how to respond to, live with, and plan for change (Greg Llyod et al., 2013). Communities need to be comprehensively involved in identifying local vulnerabilities and in the design of responses to these; from a justice perspective, meaningful inclusion in this way would meet both recognition and participatory capabilities needs (Schlosberg, 2012).

Box 7.2: Key Data Points—Shared Sustainable Prosperity
Communities need to be comprehensively involved in identifying local vulnerabilities and in the design of responses to these; from a justice perspective, meaningful inclusion in this way would meet both recognition and participatory capabilities needs (Schlosberg, 2012).

Table 7.1 Elements of community resilience building

Elements of community resilience	*Key features*
Communication	• Contextualized warnings • Clear, consistent and targeted messaging • Effective dissemination structures • Use of digestible/relatable language
Collaboration	• Sufficient space and time to focus on roles • Development of trusted relationships • Collaborators as partners • Integrated and responsive processes • Recognition as priority
Co-production	• Incorporate the needs of stakeholders • Salient decision-making processes • Large, diverse and representative range of voices • Favorable to bottom-up responses

The argument here is that capabilities form a central element of just, resilient and sustainable communities. Communities need to be facilitated with the platforms, processes and recognition through which their capabilities can be cultivated, acknowledged and importantly, used to deliver resilient sustainability. Governance models sensitive to the socio-cultural context and approaches to knowledge development which are not expert-driven or reliant would seem to be key requirements.

The Importance of Geography

Finally, fundamental geographical concepts such as scale and connection, governance and power relations, proximity and distance and space and place are fundamental to understanding processes of transition, vulnerability and resilience building. In an analysis of Blue Economy, Garland et al. (2019, pp. 13–14), identify these concepts as essential to developing a 'deeper understanding of the spatial and temporal elements of the BE (Blue Economy)'. Garland et al. (2019) go on to argue that the multi-scalar socio-ecological complexities of regions are critical to achieve a just transition. For coastal areas, for instance, management responses need to consider policy, environment and geography (Flood & Schechtman, 2014), with a candid recognition of complexity across spatial, jurisdictional and temporal scales (Heber Dunning, 2020). Adaptation responses will necessarily differ from place to place, due to diverse geographical, socio-cultural and socio-economic contexts. Successful initiatives in one place can produce different and perhaps unintended outcomes in another (Owen, 2020). Resilience building and adaptation measures should therefore be directed by the specific localized social landscapes in different communities (Owen, 2020).

However, embedding these elements in community resilience building is contingent on real capacity building, supported by and in conjunction with, the provision of financial resources (Howath, 2019). '*Navigating a just and inclusive path towards sustainable oceans will require significant will and adequate support from policy circles and funding from foundations, governments and multi-lateral funding agencies*' (Bennett, 2018, p. 143). Of particular relevance, here are questions of political and economic power, community autonomy and subsidiarity. Garland et al. (2019) discuss the importance of power structures and the role of scale on decision-making structures and processes, in particular highlighting the locus of where decisions are taken and the scope of jurisdictional and economic impact of these decisions. Subsidiarity is an important consideration here. As discussed by McGinlay et al. (2021), there is a limit to what can be devolved down to local communities in the absence of structure, guidelines, funding and supervision provided by State or Federal authorities. Coordinated planning is essential and in practice leads to higher levels of community resilience in mitigating disaster losses and enhancing adaptation (Kim et al., 2020). Links between the local scale and national and international policy scales and a broader sense of vision and direction all require definitive input from state actors (McGinlay et al., 2021). As put by Heidkamp et al. (2021, p. 11), 'a just and equitable transition needs to emerge organically and collaboratively through a "top-led and bottom-fed"[4] approach in order to assure local adaptability and compatibility'. For genuine subsidiarity to be a success, sufficient levels of trust, autonomy and resources need to afforded regional economies so that place-based sustainable economy models can evolve and prosper.

Box 7.3: Key Data Points—Shared Sustainable Prosperity

A just and equitable transition needs to emerge organically and collaboratively through a "top-led and bottom-fed" approach in order to assure local adaptability and compatibility. Heidkamp et al. (2021, p. 11)

The Challenge(s) Ahead

The carbon intensity of every dollar in the global economy needs to be reduced significantly in the years to 2030 and beyond. While the carbon intensity of an individual dollar in the global economy (expressed as kg CO_2e per PPP $ of GDP) has fallen steadily in the years since 2011, these efficiency gains have been offset by the continued growth of the overall size of the global economy. Simply put, while each individual dollar in the global economy is associated with lower carbon emissions now than in 2011, there is more carbon being emitted overall, as there are more dollars overall. As a result, there has been a continued rise in global GHG emissions, slowed in recent years only by pandemic-related lock-downs. These dynamics represent a relative decoupling of carbon emissions from

[4] Phrase coined by Kokorsch (2018).

economic growth. However, what is needed is an absolute decoupling, that is, that emissions start to fall year on year, effectively decoupled from levels of economic output. Figure 7.2 clearly illustrates the contradiction (and maybe impossibility) of pursuing both economic growth and carbon reduction public policy imperatives. The higher the growth levels and the more stringent the emissions restrictions (determined by warming targets), the lower the carbon intensity of the global economy must be. The numbers in Table 7.2 and Fig. 7.1 are stark (Morrissey & Heidkamp, 2022):

Data in Table 7.2 are presented in kg per PPP $ of GDP and were first reported in Morrissey and Heidkamp (2022). Figures in Table 7.2 can be

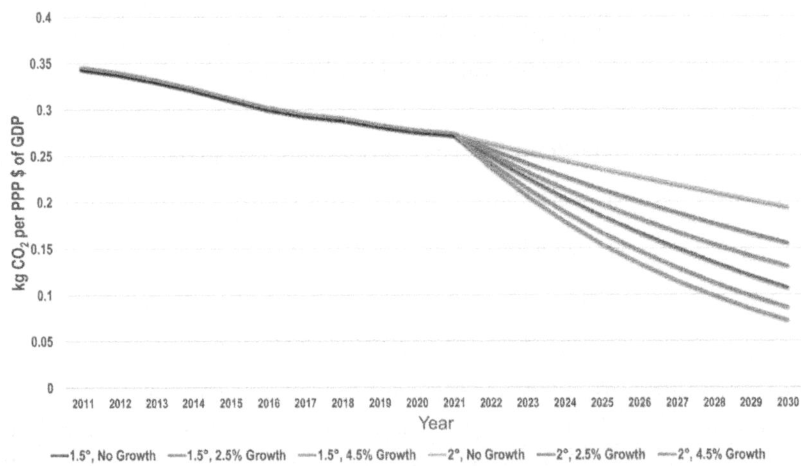

Fig. 7.2 Carbon intensity of global economy to 2030 under 6 different scenarios[5]

[5] About Fig. 7.2 (first published in Morrissey & Heidkamp, 2022): Historical data on carbon intensity for 2011–2019 from Our World in Data (2021). Global emissions data for 2020 and 2021 from Statistica (2021). Economic growth scenarios based on average global growth rate 2010 to 2020 of ~2.5% and max global growth rate of ~4.5% (in 2010) sourced from World Bank (2021). Global GDP figures for 2020 and 2021 from the International Monetary Fund (2021). Emissions cuts scenarios of 7.6% per year to meet the 1.5 °C goal and 2.7% per year for the 2 °C goal from UNEP (2019). Emissions cut figures have been adjusted to account for actual economic and emissions growth in 2020–2021, with emissions cuts 'missed' for 2021 distributed across 2022–2030. This analysis was first reported in Morrissey and Heidkamp (2022).

Table 7.2 Carbon intensity of global economy to 2030 under 6 different scenarios

	0% GDP growth	*2.5% GDP growth*	*4.5% GDP growth*
1.5 °C pathway	0.11	0.09	0.07
2 °C pathway	0.19	0.16	0.13

benchmarked against a figure of ~0.28 (kg per PPP $ of GDP), the 2019 baseline. Figure 7.2 shows the pathways for the 6 outlined scenarios to 2030 in the context of historical trends of a reduction in global carbon intensity.

A 'best of all worlds' scenario where we successfully restrict global carbon emissions sufficiently to limit warming to 1.5 °C above pre-industrial levels while also managing to maintain an annual global growth rate of 4.5% would require a reduction of carbon intensity to 0.07 (kg per PPP $ of GDP) by 2030. This represents a 75% reduction in carbon intensity in 8 years, when compared to the baseline figure of 0.28 (kg per PPP $ of GDP) for 2019.

Box 7.4: Key Data Points—Shared Sustainable Prosperity
A 'best of all worlds' scenario where we successfully restrict global carbon emissions sufficiently to limit warming to 1.5 °C above pre-industrial levels while also managing to maintain an annual global growth rate of 4.5% would require a reduction of carbon intensity to 0.07 (kg per PPP $ of GDP) by 2030—a 75% reduction compared to a 2019 baseline (Morrissey & Heidkamp, 2022).

As described by Jacobs (2021, p. 8) 'much more far-reaching interventions which will be needed if fossil fuels are to be squeezed out of the global economy and investment in green solutions increased to the levels required'.

References

Bennett, N. J. (2018). Navigating a just and inclusive path towards sustainable oceans. *Marine Policy, 97*, 139–146. https://doi.org/10.1016/j.marpol.2018.06.001

Bowles, P., MacPhail, F., & Tetreault, D. (2019). Social licence versus procedural justice: Competing narratives of (Il) legitimacy at the San Xavier mine, Mexico. *Resources Policy, 61*, 157–165. https://doi.org/10.1016/j.resourpol.2019.02.005

Brighouse, H., & Robeyns, I. (Eds.). (2010). *Measuring justice: Primary goods and capabilities*. Cambridge University Press.

Catney, P., Macgregor, S., Dobson, A., et al. (2014). Big society, little justice? Community renewable energy and the politics of localism. *Local Environment, 19*, 715–730. https://doi.org/10.1080/13549839.2013.792044

Cavaleri Gerhardinger, L., Martins de Andrade, M., Ribeiro Correa, M., & Turra, A. (2020). Crafting a sustainability transition experiment for the Brazilian blue economy. *Marine Policy, 120*. https://doi.org/10.1016/j.marpol.2020.104157

Davis, M. (2010). Who will build the ark? *New Left Review, 61*, 29–46.

Dovie, D. B. K. (2019). Case for equity between Paris Climate agreement's co-benefits and adaptation. *Science of the Total Environment, 656*, 732–739. https://doi.org/10.1016/j.scitotenv.2018.11.333

Eckstein, D., Hutfils, M., & Winges, M. (2019). *Global Climate Risk Index 2019. Who suffers most from extreme weather events? Weather-related loss events in 2017 and 1998 to 2017*. Germanwatch e.V.

Fitzgibbons, J., & Mitchell, C. L. (2019). Just urban futures? Exploring equity in "100 resilient cities". *World Development, 122*, 648–659. https://doi.org/10.1016/j.worlddev.2019.06.021

Flood, S., & Schechtman, J. (2014). The rise of resilience: Evolution of a new concept in coastal planning in Ireland and the US. *Ocean and Coastal Management, 102*, 19–31. https://doi.org/10.1016/j.ocecoaman.2014.08.015

Forsyth, T. (2014). Climate justice is not just ice. *Geoforum, 54*, 230–232. https://doi.org/10.1016/j.geoforum.2012.12.008

Forsyth, T. (2018). Is resilience to climate change socially inclusive? Investigating theories of change processes in Myanmar. *World Development, 111*, 13–26. https://doi.org/10.1016/j.worlddev.2018.06.023

Garland, M., Axon, S., & Morrissey, J., et al. (2019). The blue economy: Identifying geographic concepts and sensitivities. *Geography Compass, 1–21*. https://doi.org/10.1111/gec3.12445

Gleeson, B. J. (2018). A dangerous transition to hope. In T. Moore, F. de Haan, R. Horne, & B. J. Gleeson (Eds.), *Urban sustainability transitions* (pp. 35–49). Springer.

Greg Llyod, M., Peel, D., & Duck, R. W. (2013). Towards a social—Ecological resilience framework for coastal planning. *Land Use Policy, 30*, 925–933. https://doi.org/10.1016/j.landusepol.2012.06.012

Heber Dunning, K. (2020). Building resilience to natural hazards through coastal governance: a case study of Hurricane Harvey recovery in Gulf of Mexico communities. *Ecological Economics, 176*, 106759.https://doi.org/10.1016/j.ecolecon.2020.106759

Heidkamp, C. P., Garland, M., & Krak, L. (2021). Enacting a just and sustainable blue economy through transdisciplinary action research. *Geographical Journal, 00*, 1–13. https://doi.org/10.1111/geoj.12410

Herrfahrdt-Pähle, E., Schlüter, M., Olsson, P., et al. (2020). Sustainability transformations: Socio-political shocks as opportunities for governance transitions. *Global Environmental Change, 63*, 102097. https://doi.org/10.1016/j.gloenvcha.2020.102097

Holland, B. (2017). Procedural justice in local climate adaptation: Political capabilities and transformational change. *Environmental Politics, 26*, 391–412. https://doi.org/10.1080/09644016.2017.1287625

Howath, C. (2019). *Resilience to climate change*. Palgrave Macmillon.

Institute for Economics & Peace. (2020). *Ecological threat register 2020: Understanding ecological threats resilience and peace*. Institute for Economics & Peace.

International Monetary Fund. (2021, October). *Recovery during a pandemic*. World Economic Outlook. https://www.imf.org/en/Publications/WEO/Issues/2021/10/12/world-economic-outlook-october-2021. Accessed 20 Feb 2022.

Jacobs, M. (2021, December 15). Reflections on COP26: International diplomacy, global justice and the greening of capitalism. *The Political Quarterly*, 1–8. https://doi.org/10.1111/1467-923X.13083

Kim, H., Marcouiller, D. W., & Woosnam, K. M. (2020). Coordinated planning effort as multilevel climate governance: Insights from coastal resilience and climate adaptation. *Geoforum, 114*, 77–88. https://doi.org/10.1016/j.geoforum.2020.05.023

Kinley, R. (2017). Climate change after Paris: From turning point to transformation. *Climate Policy, 17*, 9–15. https://doi.org/10.1080/14693062.2016.1191009

Kokorsch, M. (2018). *Mapping resilience—Coastal communities in Iceland*. University of Iceland.

Lewis, N. (2018). *New blue economy in Kaikōura: A participatory process approach*. National Science Challenges Sustainable Seas. https://www.sustainableseaschallenge.co.nz/our-research/new-blue-economy-in-kaikoura-a-participatory-process-approach/. Accessed 11 Jan 2021.

McGinlay, J., Jones, N., Clark, J., & Maguire-Rajpaul, V. A. (2021). Retreating coastline, retreating government? Managing sea level rise in an age of austerity. *Ocean and Coastal Management, 204*, 105458. https://doi.org/10.1016/j.ocecoaman.2020.105458

Morrissey, J., & Heidkamp, C. P. (2022). Sustainability after Covid: Pillars for a just transition. *Environmental Sustainability*. https://doi.org/10.1007/s42398-022-00231-y

Our World in Data. (2021). CO_2 emissions and GDP. https://ourworldindata.org/grapher/co2-emissions-and-gdp?country=~OWID_WRL. Accessed 20 Feb 2022.

Owen, G. (2020). What makes climate change adaptation effective? A systematic review of the literature. *Global Environmental Change, 62*, 102071. https://doi.org/10.1016/j.gloenvcha.2020.102071

Popke, J., Curtis, S., & Gamble, D. W. (2016). A social justice framing of climate change discourse and policy: Adaptation, resilience and vulnerability in a Jamaican agricultural landscape. *Geoforum, 73*, 70–80. https://doi.org/10.1016/j.geoforum.2014.11.003

Sandel, M. J. (2009). *Justice: What's the right thing to do?* Penguin Books Ltd.

Schlosberg, D. (2012). Climate justice and capabilities: A framework for adaptation policy. *Ethics & International Affairs, 26*, 445–461.

Sen, A. (2008). Capabilities, lists, and public reason: Continuing the conversation. *Feminist Economics, 10*, 77–80. https://doi.org/10.1080/1354570042000315163

Sharifi, A. (2021). Co-benefits and synergies between urban climate change mitigation and adaptation measures: A literature review. *Science of the Total Environment, 750*, 141642. https://doi.org/10.1016/j.scitotenv.2020.141642

Statistica. (2021). *Annual CO2 emissions worldwide from 1940 to 2020*. Energy & Environment Emissions. https://www.statista.com/statistics/276629/global-co2-emissions/. Accessed 20 Feb 2022.

UNEP. (2019). *Emissions Gap Report 2019*. United Nations Environment Programme.

UNEP. (2020). *Emissions Gap Report 2020*. United Nations Environment Programme.

UNEP. (2021). *Emissions Gap Report 2021: The heat is on—A world of climate promises not yet delivered—Executive summary*. UNEP.

Williams, S., & Doyon, A. (2019). Justice in energy transitions. *Environmental Innovation and Societal Transitions, 31*, 144–153. https://doi.org/10.1016/j.eist.2018.12.001

Wilson Furlan Matos Alves, M., & Barberio Mariano, E. (2018). Climate justice and human development: A systematic literature review. *Journal of Cleaner Production, 202*, 360–375. https://doi.org/10.1016/j.jclepro.2018.08.091

Wood, N., & Roelich, K. (2019). Tensions, capabilities, and justice in climate change mitigation of fossil fuels. *Energy Research & Social Science, 52*, 114–122. https://doi.org/10.1016/j.erss.2019.02.014

World Bank. (2021). *GDP growth (annual %)*. In: Data. https://data.worldbank.org/indicator/NY.GDP.MKTP.KD.ZG. Accessed 20 Feb 2022

CHAPTER 8

Conclusions

Abstract The post-Covid-19 re-build represents a once-in-a generation opportunity to markedly shift development models to more sustainable pathways, to rebalance the domains of sustainability and in the process, to address longer-term crises including those of climate and biodiversity loss. In this volume, we have outlined 5 pillars which represent a triple bottom line framing of sustainability, of mutually supportive domains of economic, social and environmental well-being. The contribution of the pillars is in underlining the triple bottom line linked, nature our current crises and to help chart responses to start to build real prosperity in our communities. Our crises consist of ecological, social and environmental dimensions. It follows that adequate responses must also work across the domains of sustainability. The 5 pillars are informed by principles of distributive and procedural justice, recognizing the importance and advantages of real community engagement and empowerment and giving due respect and deference to the ecological carrying capacity of our fragile planet.

Keywords Development models · Prosperity · Sustainability · Economic · Social and environmental dimensions · Distributive and procedural justice · Ecological carrying capacity

J. Morrissey and C. P. Heidkamp, *Demanding Sustainability*,
https://doi.org/10.1007/978-3-031-18958-6_8

In motivating people to love and defend the natural world, an ounce of hope is worth a tonne of despair. George Monbiot (2013)

Seizing the Moment

In this volume, we have outlined 5 pillars which represent a triple bottom line framing of sustainability, of mutually supportive and equally prioritized domains of economic, social and environmental well-being and flourishing. The pillars underline the triple bottom line and relational nature of the ecological, economic and social crises which we face. In the aftermath of the Covid-19 pandemic, a number of useful overview framings of the 'crises of sustainability' have been published. However, we argue that many of these works have been too narrowly framed, frequently overlooking important aspects in part. The warnings issued by Ripple et al. (2017) and Büscher et al. (2020), for example, are welcome and represent vital interventions. However, while Ripple et al. (2017) provide some excellent strategies which would serve to inform our 'pillar one', their warning pays insufficient attention to social, justice and governance aspects. Similarly, Büscher et al.'s (2020, p. 1) five key policy proposals for a post-Covid-19 development model are useful and focus strongly on the key issues of consumption, redistribution and debt cancellation. However, their proposals do not sufficiently emphasize the need for rapid decarbonization of the global economy, the urgency of preserving life-supporting ecosystems or the need for strong governance reforms to underpin successful future transition. These publications also do not explicitly address the need to face up to the ever-increasing challenges of adaptation and the imperative for just responses to climate disruptions, framed here as pillar five: a just resilience. In this volume we have placed the principles of distributive and procedural justice as central to the pillars. We have also emphasized the importance and advantages of real community engagement and empowerment. A strong sustainability framing means that the ecological carrying capacity of our fragile planet is presented in our foundational pillar one, an ecological prosperity.

A number of tensions are evident on critical reflection on the practical reality of implementing these pillars. Some fundamental questions remain: how can global poverty be reduced in a dematerializing system, characterized by reduced consumption and production? How can progress towards the sustainable development goals be sustained in a world of ecological breakdown and climate disruption crises? How can we assure a transition to a more just and equitable society under these conditions?

The authors recognize that a 'governance new deal' based on bottom-up participatory models, as proposed in pillar four: a transformative social sustainability, is contingent on an unfolding crisis that remains on some level 'manageable'. The case for strong top-down intervention remains obvious, to set'rules of the game', to fund investment, to guide planning and perhaps above all, to coordinate and align responses across space, sectors and scales. Our argument is that such co-ordination should facilitate and empower rather than impede grassroots participatory responses. In the event of a run-away crisis, the appropriate governance response may better align with Gleeson's (2018) 'Guardian State', committed to collective welfare in a time of 'shared endangerment'. Such a vista, however, would entail complete disruption of social and economic systems where analysis from a 2022 perspective would likely be entirely irrelevant in any case.

Bouncing Forward

While we strongly agree with the calls of Ripple et al. (2017), Büscher et al. (2020) and Swilling and Annecke (2012) amongst others for reduced consumption in the advanced economies, there remains a very great risk and strong likelihood that the end of Covid-19 pandemic restrictions will see a 'bounce back' of increasing global consumption trends. Citizens subjected to months of lock-down conditions will likely wish to resume 'normal' life at the first opportunity. While in this volume we have framed this moment as a once-in-a generation opportunity to decidedly shift development patterns to more sustainable pathways, there is a danger that the wrong lessons will be drawn from the pandemic, that a lifting of pandemic restrictions will serve to further fuel accelerated materialist consumption lifestyles. Internationally, a 'green' pandemic recovery could reduce further the emissions reductions to be achieved by 2030 through implementation of unconditional NDCs (UNEP, 2020).

As the policy focus shifts from a mitigation perspective to one of adaptation in the coming decades, questions of justice, just transition and indeed just disruption may become even more pertinent. In the context of adaptation, for instance, how will issues of strategic retreat, the abandonment of settlements and infrastructure or radical shifts away from certain economic sectors be framed in procedural and distributive justice terms?

In the context of limited public funds or public funds that are under pressure as a result of climate-related costs, how will considerations of climate adaptation impact on vulnerable groups?

In framing just transitions, the classic human geography themes of space, place, scale as well as power relations are central considerations. Space and place necessarily dictate current processes of uneven development as well as future prospects for transition/green development. There is considerable risk that geography will determine socio-spatial differentiation in transitions processes. We must acknowledge that transition is unlikely to happen in a spatially homogenous fashion. Local characteristics including capacity limitations, issues of power asymmetry as well as cultural and institutional contexts may all have important roles to play in the successful delivery of transition or not. The transitions literature has addressed issues of complexity, path dependency and inertia in socio-technical systems change in some detail to date. However, the implications of these issues from resource distribution, fairness and social justice perspectives remains understudied. What is obvious is that vulnerability is likely to increase for many groups in the coming decades. As well as the obvious exposure to climate-related hazards, including for instance storm surges, flooding, drought, etc., low-income groups may find themselves to be highly exposed in the context of radical policy change or in the context of radical economic restructuring as our economies decarbonize. Attention to scale and importantly to scalar-mismatches in prescribed solutions are therefore important.

Getting the Job Done

Policy responses need to be 'laser-focused' on practical, realizable and cost-effective responses. As clearly presented in the Project Drawdown work, the technology is at hand to deal with the climate crisis (Hawken, 2017; Wilkinson et al., 2020). The challenge is one of scale and investment. What is critical is that in the limited timeframe available to proactively address the crisis, we collectively, as a global community, use our resources wisely. It is critical that scarce resources are used to maximum impact in achieving carbon emissions reductions and in reducing community vulnerability. The type of cost–benefit analysis conducted by Project Drawdown is a useful starting point, but this type of analysis needs to be scaled up, intensified and applied across scales from global to national to regional and local. As the climate crisis and policy responses to it become increasingly politicized in the coming years and decades, the importance of an evidence-based approach to policy development becomes ever more pressing. Responses need to tackle the actual problem, as opposed to be seen to be tackling the problem. In this context, tried and tested ideas, approaches and technologies incur a premium for at-scale deployment, in contrast to untested radical innovations. Niche innovations from the previous three decades of sustainable development experimentation should be re-examined. We argue that this is best done from a transdisciplinary perspective (Nicolescu, 2008) rooted in critical pragmatism (Forester, 1993) which is focused on co-learning amongst stakeholders in government, business, academia and the NGO community. As academics, it behooves us to champion this open dialogue like never before.

References

Büscher, B., Spierenburg, M., Bavinck, M., et al. (2020). *Planning for post-corona: Five proposals to craft a radically more sustainable and equal world*. Amsterdam.

Forester, J. (1993). *Critical theory, public policy, and planning practice: Toward critical pragmatism*. SUNY Press.

Gleeson, B. J. (2018). A dangerous transition to hope. In T. Moore, F. de Haan, R. Horne, & B. J. Gleeson (Eds.), *Urban sustainability transitions* (pp. 35–49). Springer.

Hawken, P. (2017). *Drawdown: The most comprehensive plan ever proposed to reverse global warming*. Penguin.

Monbiot, G. (2013). For more wonder, rewild the world. In *TEDGlobal 2013*. https://www.ted.com/talks/george_monbiot_for_more_wonder_rewild_the_world/transcript?share=1ca49816c2. Accessed 1 Dec 2021.

Nicolescu, B. (2008). *Transdisciplinarity—Theory and practice*. Hampton Press.

Ripple, W. J., Wolf, C., Newsome, T. M., et al. (2017). World scientists' warning to humanity: A second notice. *BioScience, 67*, 1026–1028.

Swilling, M., & Annecke, E. (2012). *Just transitions: Explorations of sustainability in an unfair world*. United Nations University Press.

UNEP. (2020). *Emissions Gap Report 2020*. United Nations Environment Programme.

Wilkinson, K., Chissell, C., & Foley, J. (2020). *The Drawdown review: Climate solutions for a new decade*. Project Drawdown.

Index

J. Morrissey and C. P. Heidkamp, *Demanding Sustainability*,
https://doi.org/10.1007/978-3-031-18958-6

The manufacturer's authorised representative in the EU is Springer Nature Customer Service Centre GmbH, Europaplatz 3, 69115 Heidelberg, Germany. If you have any concerns regarding our products, please contact ProductSafety@springernature.com

Printed and bound by CPI Group (UK) Ltd, Croydon, CR0 4YY
28/09/2026
02224072-0017